Speaking Statistics

Cecily Kramer Bodnar

Globe Fearon
Upper Saddle River, New Jersey 07458

Library of Congress Cataloging-in-Publication Data

Bodnar, Cecily Kramer.
  Speaking Statistics / Cecily Kramer Bodnar.
    p. cm.
  Includes index.
  ISBN 0-13-030180-9
  1. Statistics. I. Title.
QA276.12.B63 1995
658.4'033—dc20   95-45
                                                            CIP

*To the women and men
who work every day
in American business and industry.
They taught me
how to learn and how to teach.
Many of them are
among my personal heroes.*

Executive editor: Mark Moscowitz
Project director: Mary McGarry
Editorial/production supervision and interior design:
   WordCrafters Editorial Services Inc.
Cover design: Marianne Frasco
Manufacturing buyer: Ed O'Dougherty

 © 1995 by Cambridge Adult Education
Upper Saddle River, New Jersey 07458

All rights reserved. No part of this book may be reproduced, in any form or by any means, without permission in writing from the publisher.

Printed in the United States of America

10 9 8 7 6 5 4 3 2 1

Prentice-Hall International (UK) Limited, *London*
Prentice-Hall of Australia Pty. Limited, *Sydney*
Prentice-Hall Canada Inc., *Toronto*
Prentice-Hall Hispanoamericana, S.A., *Mexico*
Prentice-Hall of India Private Limited, *New Delhi*
Prentice-Hall of Japan, Inc., *Tokyo*
Simon & Schuster Asia Pte. Ltd., *Singapore*
Editora Prentice-Hall do Brasil, Ltda., *Rio de Janeiro*

Contents

TO THE STUDENT   *vii*

1   THE BEST BELT COMPANY   1

Welcome to the Best Belt Company, 1
Teams at Work, 3
Part 1. Evaluating the Problem, 5
    Step 1: Identifying the Problem, 5
    Step 2: Getting Ideas (Brainstorming), 5
    Step 3: Grouping Ideas, 6
    Step 4: Choosing the Populations, 6
Summary, 8
Activities, 9

2   BEGINNING TO SOLVE A PROBLEM   13

The Problem-Solving Process, 13
Gathering Good Information, 14
    Sampling, 14
    Collecting Data, 15
Analyzing the Data, 17
    Using Table Data, 17
    Frequency Distributions, 18
Predicting the Cause, 19
Summary, 19
Activities, 19

3 NO EASY ANSWER 23

Evaluating the Change, 23
    Score It!, 23
Question It Again, 24
    Cause and Effect or Coincidence?, 24
    Choosing a Sample, 25
    Gathering the Data, 26
Setting Up a Tally Chart, 26
    Cell Size and Boundaries, 28
    Setting Up Cells, 30
Summary, 33
Activities, 33

4 MEASURES OF THE MIDDLE 37

Tolerances, 37
Central Tendency, 40
    The Median, 41
    The Mode, 42
Plotting a Normal Distribution, 42
Summary, 44
Activities, 45

5 SOLVING THE PROBLEM 47

Reading a Line Graph, 47
Dispersion or Spread, 49
Standard Deviation, 52
Probability in the Normal Distribution, 53
Using Estimates, 54
Comparing Curves, 54
Using the Data to Find a Reason, 55
Summary, 57
Activities, 60

6 USING LINE GRAPHS 63

Scattergrams, 65
    Moving from Scatter to Line, 66
    Changing the Scale on Line Graphs, 68
Cycles, 70
    Finding a Cycle, 70
    The Meaning of Cycles, 71
The Importance of Scale and Interval, 72
Run Charts, 72
Graphing Deviation from the Range Midpoint, 73
Summary, 75
Activities, 75

CONTENTS

7 USING BAR GRAPHS 77
Using Bar Graphs to Show Distributions, 81
Bimodal Distributions, 84
Summary, 86
Activities, 87

8 USING CIRCLE OR PIE GRAPHS 89
Determining Percentage of Total, 90
Circle Graphs, 91
    Making a Circle Graph, 92
    Comparing Circle Graphs, 94
Summary, 101
Activities, 101

APPENDIX A ANSWERS AND ASSISTANCE 103

APPENDIX B QUICK MATH REVIEW 115
Whole Numbers, 115
    Place Value, 115
    Signed Numbers, 119
Measuring in Metrics, 121
Fractions, 123
Decimals and Percents, 125

GLOSSARY 131

INDEX 135

To the Student

Welcome to *Speaking Statistics*. This book was written for you and for the men and women who learned with us in a learning lab setting of a major manufacturing firm. Our partners in learning were using practical statistical process and quality control on the production lines. They were measuring and analyzing their own productivity as building services, distribution, and office support personnel. *Speaking Statistics* is the result of their need to understand the statistical process and the standards set by their industry.

This book will help you understand why certain kinds of information are important. The team in the Best Belt Company has problems to solve, just as you do. They know that accurate information is critical to making good problem-solving choices. The team uses ordinary math skills: adding, subtracting, multiplying, and dividing to work with the information they gather. There are no mysterious formulas here.

As you work your way through the book, you will find many activities to help you practice what you are learning. You are encouraged to find activities in your own workplace. Combine your work and learning. Ask questions. Every question is worth asking if it helps you understand how what you do helps the whole company.

Speaking Statistics is set in a story, because stories help us to remember. As you apply what you learn, ask yourself, "What would Connie or Roy do next?" In Chapter 5, you will find a problem-solving pattern you can use often in your work or learning. It will help you remember to do all the steps.

Speaking Statistics uses many words we don't see every day. Each chapter has sections called "Words at Work." These sections explain the terms in words you know. Don't be afraid to ask if the meaning is not clear to you. It is important to understand this vocabulary. You may already be hearing the "Words at Work" on your job or in your learning program.

Use the answer key in Appendix A to check your answers. If you're still not clear, ask your instructor to explain.

There are many interesting formulas in statistics. If you like to work with numbers, you may want to know more about them. Ask your instructor or your supervisor, or look at your local library for an introduction to statistics for business.

As you use the book, keep in mind that statistics are ways to tell the story of what is happening. They are always estimates—sometimes we call them "educated guesses." Statistics always tell about groups of events or things; they really do not tell much about any *one* event or thing. So, statistics are useful as the first step in problem solving. They rarely solve the problem by themselves. It takes women and men like you to do that.

The folks at Best Belt Company are eager to help you join them in *Speaking Statistics*.

Acknowledgments

The author would like to thank everyone who believed that I really could write this book. Don Kramer, who gave me a life-long love for words; Alex Bodnar, who held our world together while I was otherwise engaged; Jim Brown and Mark Moscowitz for the opportunity to write it; and especially Bob McIlwaine, who encouraged, sustained, and stayed the course—all the while in the midst of a tidal wave of change.

Cecily Kramer Bodnar

1

The Best Belt Company

■ ■ ■ ■ ■ **GOAL**
To learn how teams begin to work on quality and process control.

WORDS AT WORK

> There are many words that have special meanings in quality and process control. You will find them in color-tinted boxes throughout the book. If you already know the words, just keep reading. If the words are new to you, take a minute to read about them. Special words used in this chapter are: **cross-functional team**, **quality**, **process**, **brainstorm**, **data**, **population**, and **specification**.

WELCOME TO THE BEST BELT COMPANY

In this chapter you will meet some key people at Best Belt. They have been asked by their manager to form a team. They will help each other solve problems. You will be a team member, too. The team wants to make Best Belt live up to its name. They want it to be the best place to buy machine belts.

Best Belt makes drive belts that run machines. The machines are used in homes and businesses. Some of the machines are floor polishers, vacuum cleaners, sewing machines, washers, and dryers. Best Belt sells its products to companies that make and repair these machines. Other companies make the same kind of belts. They are always trying to take sales away from Best Belt.

George Lindsey is the plant manager. His job is to be sure everyone works together to keep sales growing. George supervises five departments. The Purchasing Department buys the materials and supplies to make the belts. Maria Torres is purchasing manager. Her job is to work with suppliers to obtain the best quality materials at the lowest price. She must also be sure the materials get to the factory on time. The Manufacturing Department takes the material and makes long tubes. There are many different sizes of tubes—one for each kind of belt. The Manufacturing Department also cuts each tube into belts. Roy Urumatsu is the manufacturing manager. Roy's job is to be sure each belt is the right size and strength. Roy and Terry Ndele work together often. Terry manages the Building and Maintenance Department. Part of his job is to be sure that all Roy's machines are working. Terry

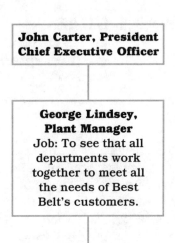

FIGURE 1.1
The Best Belt Company organization chart

checks and fixes the machines on a schedule. Roy also calls Terry whenever there is a machine problem.

The Marketing Department sells the belts to companies where the machines that use belts are made or repaired. Marketing also advertises the products of Best Belt. Connie Borrelli is the marketing manager. Connie wants her customers to come to Best Belt whenever they need belts. She works closely with Peter Stanek who is the manager of the Customer Service Department. Peter's job is to help customers with questions or problems. He works with customers while they are using Best Belts in their machines. He also helps customers with their questions when they are repairing machines. Peter and Connie work together to be sure the customers have the right belts and know how to use them.

TEAMS AT WORK

Think about how a team works. In sports, in music (a band is a musical team), in work and family life, we often are part of a team. People who do the same job work together to share ideas and work loads. For example, four people are carrying a large log. Each person is needed, and each is doing the same job. They are working together as a team. Another kind of team could paint a house. One person could scrape off the old paint. Another could paint the window and door trim. A third person could paint a wall. They are each doing a part of the work of painting the house.

There can be many teams in manufacturing. Moving the material, running a machine, sorting, and packing are all jobs of the team manufacturing a product. These teams are work teams. Members are usually from the same department. They are working together to do one job.

The Best Belt Company is beginning a new kind of team. The people on this team do not have the same job. They do not work in the same department. Their goal is to bring different departments together to solve problems they cannot solve alone. The problems are too large for one department to handle.

So, the Best Belt Company has begun a **cross-functional team.** This kind of team can solve problems because, together, they know about the whole business. Suppose a problem with a belt comes up because the belt slips when the machine is running fast. The problem might not show up until Connie sells the belt. The person who buys the belt calls Peter and tells him the belt is not working. How will Peter know what the trouble is?

THE BEST BELT COMPANY

WORDS AT WORK

> **cross-functional team** (kross-**funk** shun al) People who have jobs in different departments coming together to work on a plan or solve a problem. The problem might involve more than one department.

Peter will ask the customer some questions about the belt. He will probably call Connie to be sure the customer has the right belt. He might call Roy to ask about problems with making the belt. Peter's staff cannot fix this problem alone. Peter needs the help of the cross-functional team. Together, they will be able to check into every part of the business.

The goal of the team is to make sure that every job at Best Belt is done perfectly. It is a quality management team. Best Belt's managers expect to make all their customers happy. They will help each other in many ways. They will set up a process for each department. Then they will check on each process regularly. They will look at how each department has to cooperate with all the others.

If Roy's department, Manufacturing, does not make the belts well, Connie's salespeople will soon not be able to sell them. Peter's department, Customer Service, cannot explain how to use a belt that is not made right. It will not work well no matter how it is used. That is what quality teams are all about. Their job is to keep the whole company running as perfectly as possible—to maintain total **quality**.

Part of your training is to observe how a quality team works. You will help get information. The team will be looking at the whole **process**.

WORDS AT WORK

> **quality** (**kwa** li tee) A state of excellence.
> **process** (**pro** ses) A way of doing something. Mixing ground meat with salt and pepper, cooking it, and putting it in a bun is a process for making hamburgers.

At Best Belt, the Quality Team meets every Monday afternoon to share information about new products or ideas. They also share problems and help each other. At this week's meeting, Peter says he has a problem.

THE BEST BELT COMPANY

Part 1. EVALUATING THE PROBLEM

Step 1: Identifying the Problem

The Customer Service Department at Best Belt is getting some complaints from floor polisher customers. The belts made for the new floor polisher model are breaking too soon. The belts are expensive to replace. Also, the polisher cannot be used until the new belt is put in. The customers are not happy with Best Belt. "I need your help to find out what's happening to these belts," Peter says. "Does anyone have an idea?"

"Are you sure they're using the belts we made for that model, #295?" asks Roy.

"I'm sure that's the one my salespeople sold to be used in the polisher," Connie answers. "It's the only one that will fit the new model."

"Let's **brainstorm** to see if we can come up with some ideas," George suggests.

WORDS AT WORK

> **brainstorm** (**brain** storm) A way to get many ideas quickly. Team members offer ideas as fast as they can think of them. Every idea is okay and written down until there are no more. Ideas can be grouped and then discussed.

Step 2: Getting Ideas (Brainstorming)

"Good!" says Maria. "I'll write down our ideas. Remember the rules. Say whatever you think of. Do not say if anyone's idea is good or bad. Keep going until we run out of ideas. Who has an idea to share?"

Maria writes down these ideas from the team:

Wrong belt.
Not put in machine correctly.
Too thick.
Too thin.
Too narrow.
Too wide.
Workers not trained enough.
Cutter machine broken.
Are some days' belts better than others?
How many break? Is this a big problem?
How many should break out of a day's output?
What shift has the most belts break? Can we tell?

"Well, we raised a lot of questions," George says. "I would say we need some information before we can go on."

THE BEST BELT COMPANY

Step 3: Grouping Ideas

"First, let's put the ideas we got into some order," says Connie. "We have questions about

1. The size of belts,
2. The workers,
3. Whether the customers know how to use the belts, and
4. The quality of belt we make.

I see four places we should look for information."

Practice: Brainstorm with your group. (You can also brainstorm alone with pencil and paper.) See how many ideas you can come up with to fix the parking problem where you live or work. Look back at the rules for brainstorming Maria gives before you begin.

Then, group your ideas. What questions can you ask about your ideas? (For example, do they cost money?) Find at least three questions you can ask.

Step 4: Choosing the Populations

The questions the team thought of are all good questions. To answer them, the team needs **data**. They need correct information about the four **populations** that must be understood to solve the problem.

WORDS AT WORK

> **data** (**day** ta) Pieces of information about a product or process.
> **population** (pop u **la** shun) A group of people or things that can be counted or measured. All the people at a ball game make up the population of the stadium.

Roy needs to know about the polisher belts. In fact, each kind of belt can be called a population. For example, Roy could take 10 belts from each shift (30 belts) and wear them out on a machine. The 30 belts will be a population.

Practice: Write down five populations you can see around you in your group, your family, or your job. (Cousins might be a family population.) How do you know who or what belongs in each population? Make a list of things that members of one population share. (Cousins are all related by blood or marriage.)

"Yes," says Roy. "I need data to tell if the belts are good quality. How many are likely to break even if we do everything right? I need to know how thick and strong good belts are."

"We all need good information. I need to know exactly what each size and kind of belt can do. That way my department can always sell customers the right belt for their machine," adds Connie.

"I will check my population of cutter machines to be sure they are working," Terry adds. "Roy, do we have any new workers who might not know how to use the machines?"

"I'll check," says Roy. "I'll also look up the engineer's design for the #295 belt. That way we'll know what good belts look like. Peter, please make sure the customer has put the belt on the machine the right way."

Practice: List the population each team member will find out information about.

Connie has several populations because she sells to different customers. Each customer has different needs. Her sewing machine manufacturers need to buy more belts at one time than the repairers do. That is because, hopefully, more machines are made than need to be fixed. So, machine makers are one population and repairers are another.

Maria's population includes suppliers of each type of belt material. Another of the Purchasing Department's populations is the shippers who bring the materials.

Peter's Customer Service Department has several populations. He wants to know about complaints as one type of population. For example, late shipments to customers and mistakes in customers' bills are two populations Peter cares about.

Right now, the Quality Team is worried about one of Roy's populations: the #295 floor polisher belt. Like many other populations, Roy's has a large number of things in it. Each week, there are about 5,000 polisher belts made. Each one of the belts is an individual. Connie might have 100 polishing machine makers in her population. Each one of the machine makers is an individual. Figure 1.2 shows individuals in a population.

Practice: How could you find out the following?

1. On which days of the week are most workers absent?
2. How many of your 500 employees prefer to have Columbus Day or Veterans' Day as a paid holiday?
3. The cafeteria serves soup and salad and hamburgers. How can you find out how many people bought soup and salad and how many bought hamburgers?

List each population you could look at. Who are the individuals in it? (Remember, not all populations are people.)

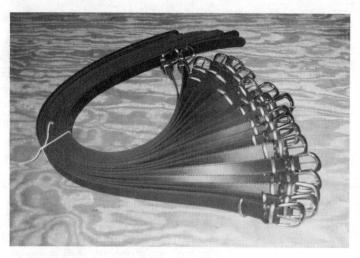

FIGURE 1.2
This population is made up of 20 individual belts. (© Jeff Taxier/ BROWN BELTS INC.)

Brainstorm at least three ways you could find out the answer to each question. (Be creative—your process doesn't have to be practical or even possible!)

WORDS AT WORK

> **specification** (spes if i **ka** shun)   Data that tell how an item should look or be. Size, strength, color, and material are just some pieces of information that might be called specifications or "specs."

Roy can tell which belts are in this population. Each population can be described in some way. For example, part #295, floor polisher belts, should all be the same length, thickness, and strength. The data that describe a population of things may be called **specifications** or "specs." When Roy says he needs to look at the belts to see how thick and strong they are, he is talking about specifications. In the next chapter, the team will decide how to help Roy check his population of polisher belts.

SUMMARY   You have been asked to join the Quality Team at Best Belt. The members include George Lindsey, Maria Torres, Peter Stanek, Connie Borrelli, Terry Ndele, and Roy Urumatsu. Use the organization chart, Figure 1.1, whenever you want to check on their jobs. George is the plant manager. The other five are department managers who report to George. They are a cross-functional team. Each one has a different job in a different department. They all have a common goal: to make Best Belt the highest quality manufacturer of machine belts in the world.

Best Belt's cross-functional Quality Team members carried out the four steps that make up part one of problem solving: 1. They identified the problem (the broken belts). 2. They then brainstormed possible causes and some questions to get data about. 3. Next, they put the ideas and questions into groups. 4. In identifying populations, they decided what groups of people and things they need to work with. Those four steps make up Part One (Evaluating the Problem) of problem solving. We will learn the other parts in the next chapters. Now they will begin to look for information to answer their questions. Roy will check the engineers' specifications. Terry will check the machines. Peter will ask customers about how they are using the belts. Connie will check that all the belts shipped to polisher customers are belt #295. Team members are getting data about their departments' contributions to belt #295.

ACTIVITIES

Activities are ways you can be sure you understand the information in the chapter. The answers and additional assistance are in Appendix A.

1. Match the best meaning with the term.

   a. team
   b. data
   c. population
   d. process
   e. quality
   f. brainstorm
   g. individual
   h. cross-functional team
   i. specification

   the best possible condition
   a group of people working together with a common goal
   information that can be found and measured or counted
   the total number of people or things
   the way things are made, moved, or used, or the way people are served
   a piece of information describing something
   different teams, departments, or jobs working together to solve problems
   a way to get many ideas quickly
   one of a population of people or things

THE BEST BELT COMPANY

2. Read the sentences below. Identify which teams are work teams (w) and which are cross-functional teams (c).
   a. Every month leaders of the Packing Department's three shifts meet to discuss getting information about problems from one shift to the next one.
   b. Managers of the Customer Service Department meet to discuss a new kind of machine belt.
   c. The managers of Marketing, Customer Service, Research, and Accounting meet to discuss when to begin making and selling a new product.
   d. The Best Belt Community Service Committee meets to plan a fundraiser.
   e. Members of the painting crew, machine maintenance team, manufacturing shifts, and plant security office meet to plan the best time to repair the manufacturing area of a plant.

3. There are three columns labeled Population, Individual, and Data. Place each term in the set into the correct column. The first set is done for you.

Population	Individual	Data
machine operators	cutters	belts cut by each worker

   a. machine operators—cutters—belts cut by each worker
   b. electricians—skilled trades—repair jobs completed
   c. gallons sprayed per minute—hand-held tanks—plant fire extinguishers
   d. Dressmaker Deluxe—sewing machines—size of belt
   e. how many people supervised—managers—Roy Urumatsu

4. If you were on a cross-functional team at your workplace, who would likely be on it with you?

5. Describe two problems a cross-functional team could work on in your workplace. You can use these problems as examples as you work through this book.

2

Beginning to Solve a Problem

■ ■ ■ ■ ■ *GOAL*
To gather and analyze data in order to solve a problem.

WORDS AT WORK

Here are the special words that appear in this chapter. Be sure to read the boxes to learn their meanings: **quality control, sample, table, range, significant, frequency, distribution.**

THE PROBLEM-SOLVING PROCESS

The Quality Team uses a five-part problem-solving process. The five parts are as follows:

1. See It: Evaluating the problem.

In Chapter 1, you learned that this part has 4 steps. Step 1 is Identifying the Problem. Step 2 is Getting Ideas or Brainstorming. Step 3 is Grouping Ideas, and Step 4 is Choosing Populations.

Brainstorming is a way of getting ideas quickly. The team thought of many ideas. They divided them into groups: belt size, workers, customer use, and quality of belt. Then, they chose to get information about the size and quality of the belts. Polisher belts are the population they will work with.

■ 13 ■

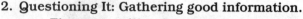

2. Questioning It: Gathering good information.
   The team will ask questions and collect the answers. They will need data from all shifts for a period of time. Part two often involves sampling.

3. Think about It: Analyzing the data.
   The team will sort the information. They will see how each piece of data fits with the others. They will look for patterns.

4. Aim at It: Predicting the cause.
   The team will choose the most likely cause and make a change in the process.

5. Score It: Evaluating the change.
   The team will gather data again to be sure the problem has been solved. If it has not, they will go back to part one.

These five steps can be used to solve any problem that is not caused by blind chance or luck.

GATHERING GOOD INFORMATION

Question It

The Best Belt Company gathers data all the time. Roy wants to be sure the quality of the belts is the best it can be. He wants to control the quality of his department's work. Information about **quality control** is very important to him. If the data are in order, the team may see a pattern. Let's find out.

WORDS AT WORK

> **quality control** (**kwa** li tee kon **trol**) To check that every product meets a certain degree of excellence.

George says, "I think we should look at the plant's data on this belt. What information can we get? We need to know how many belts are breaking and why. It is too expensive to check every belt. Someone must be paid to do the checking. It will take too much time. Of course, every belt cannot be tested to see when it wears out. There will be no belts left to sell; they will all be worn out!"

Sampling

One way to get a good idea of how long the belts last is to wear out a few belts. Roy could pick some belts made at different times and see how long each lasted. This is called taking a **sample** or sampling.

WORDS AT WORK

sample A group of people or things taken out of the whole population. The sample group must be like the population in important ways. The sample individuals might be the same size, age, or fabric as the population as a whole. Studying the sample can tell about the population without inspecting every individual.

Some belts might last three months, some four, and some six with normal use. For sampling to work, some things have to be decided before the sample is taken. The sampler must be familiar with the belts' specifications. Suppose you want to know which shift makes the longest-lasting belts. (1.) You must get belt samples from all the shifts. (2.) You will want to keep each shift's samples in its own group. Then you can compare shift A, shift B, and shift C. (3.) You will also want to keep the samples in time order. That is important. Often, the time something happens will give a clue as to why it happens.

Collecting Data The team decides to check how often the polisher belts break in the plant. On each shift, a sample of belts is checked. They are put on a machine that tests how long they last. Roy suggests that the team look at a week's tests for each shift. He asks you to get the number of belts that broke too soon from each shift leader. Because the team is counting the number of belts, they are collecting countable data.

On the A shift Frank Martin counted 1 broken belt on Monday, 4 on Tuesday, 2 on Wednesday, 8 on Thursday, and eight on Friday. Ang-Sun Cole, on the B shift, found 2 on Monday, 3 on Tuesday, 3 on Wednesday, 7 on Thursday, and 8 on Friday. On the C shift, Judy Terrell had 2 broken belts on Monday, 3 on Tuesday, 4 on Wednesday, 6 on Thursday, and 9 on Friday.

How would you organize the information? Do you think it would help to put it into the form of a **table**?

WORDS AT WORK

table A way of showing data. The numbers are put in columns and rows so they can be seen quickly and clearly.

The table might look like the one in Figure 2.1. Tables need a name or title. They also need labels for the columns and rows to help you read the table quickly and clearly. In set-

BELTS BROKEN DURING TESTING

	Monday	Tuesday	Wednesday	Thursday	Friday	Week Total
A shift	1	4	2	8	8	23
B shift	2	3	3	7	8	23
C shift	2	3	4	6	9	24
Day's total	5	10	9	21	25	70

FIGURE 2.1
Organizing information in a table

ting up a table, first, you have to decide which data will be shown in rows across. The names of these groups of data are put in the column at the left. The names across the top of the chart name the columns of data.

Practice: Put the following sets of data in table form. The first table is set up for you.

EMPLOYEE ABSENCES

	Monday	Tuesday	Wednesday	Thursday	Friday	Total
Week 1						
Week 2						
Week 3						
Total						

1. On the five days of each week, these absences were recorded: week 1: 10, 13, 15, 9, 11; week 2: 11, 9, 15, 13, 16; week 3: 25, 10, 3, 6, 19. Put this data on the following table.

   a. On which day of the week are the most employees absent?

   b. What is the total of absences for each day? For each week?

2. Put the results of this vacation-day survey in table form. Decide how to label the side and top of your

table. On the A shift, 40 people wanted off on Veteran's Day and 30 wanted Columbus Day. On the B Shift, 20 wanted Veteran's Day and 50 wanted Columbus Day. On the C shift, 15 wanted Veteran's Day and 55 wanted Columbus Day.

   a. How many people on each shift made up the sample?
   b. Which holiday do you think the workers got? Why?

ANALYZING THE DATA

Using Table Data

Do you think it is easier to study the data when it is in table form? The team could see that there were many more belts broken on Thursday and Friday than on the other days. On Monday, only 5 belts broke. On Friday, 25 belts broke. When you subtract the least, 5, from the greatest, 25, you get the **range**, 20. The range shows the difference between the best day, Monday, and the worst day, Friday. There is an important difference between Monday and Friday. It is **significant** that 70 belts broke and that the range was 20.

WORDS AT WORK

> **range** (raynj) A number showing how much difference there is between the smallest and the largest sample measurement.
>
> **significant** (sig **nif** i cant) Important to what we are doing. A significant thing is a "sign" that we want to pay attention to.

Significant has a special meaning in quality control. If something is significant, it happens for a controllable reason. Things that happen for a reason can usually be controlled if the reason is known. Things that happen by luck or chance cannot be controlled. It is not possible to know why they happen just where or how they do. If one belt breaks in the whole week, it might be by chance or "bad luck." It might not mean anything is wrong with the process.

On the other hand, 70 belts breaking is not likely to be just bad luck. The team is pretty sure there is a reason. Once they know the reason, they can do something about it. In this case the range is significant—it means something.

Look at the range in the week's total among the shifts. To find the range, subtract the smallest number from the largest. Shift A had 23 broken belts for the week. Shift B also

BEGINNING TO SOLVE A PROBLEM

had 23. Shift C had 24 broken belts. Can you figure the range of shift totals? That's right, the range is only 1 (24 − 23 = 1).

Is there an important difference among the shifts in the Best Belt sample? Would it be true to say that one shift is much better than the others? The wider (bigger) the range, the more significant the difference is. The range is one way of looking at a set of numbers. The team will also use other ways. One of them is frequency distribution.

Practice: (1.) Find the range of workers absent during the sample period on page 16.

(2.) Find the range of votes for Columbus Day across the three shifts in the sample on page (17).

Frequency Distributions

The information gathered by the team is a frequency distribution. The **frequency** of belts broken too soon is 70. They are distributed across three shifts and over five days. The **distribution** shows that more belts broke at the end of the week, but that broken belts were distributed quite evenly among the shifts.

WORDS AT WORK

> **frequency** (**free** kwen see) How many things there are or how often something happens. For example, a manager might want to know the frequency of sick days taken in December.
> **distribution** (dis tri **bu** shun) How the things or events are arranged. For example, the manager might want to know how December sick days are distributed. He wants to know if more sick days are taken on Mondays and Fridays than on Wednesdays. A frequency distribution table will tell him two things: how many sick days were taken and where in the work week.

The team now has an important piece of information. They know that belts made on Thursday or Friday seem more likely to break. Monday's and Wednesday's belts break the least. What they need to know is why this happens. Getting true information is a very important step in problem solving. It is a good idea to search for all the possible reasons you can find. Roy asks you to go with him to get the shift leaders' ideas.

Practice: List the frequency of absences for each day using the sample on page 16. How were the absences distributed?

a. More at the beginning of the week?
b. More at the end of the week?
c. More in the middle of the week?
d. More at both ends of the week?

PREDICTING THE CAUSE

Aim at It

Ang-Sun Cole, shift leader on the B shift, says the machines are checked and set during the weekend. They might not be working right by the time Thursday comes. She suggests checking the machines on Wednesday as well as on the weekend. Terry Ndele, head of the Maintenance Department, agrees to check the machines on Wednesday. He will check them at the end of the B shift. If machine-set is the cause of the problem, about the same number of belts should break each day of the week after they are checked. There should not be a significant difference between Monday and Friday.

In Chapter 3, the team will find out if they have the solution to the problem.

SUMMARY

Part one of the problem-solving process is to See It: Evaluate the problem.

Part two is to Question It: Gather information.

Part three is to Think about It: Analyze the data. Once the data is gathered, it is put into a frequency distribution table or graph. Making a table or graph makes it easier to see what is happening. Arranging the broken belt data in time order helped the team to see what was happening when. There were more broken belts at the end of the week. There were about the same number of broken belts on each shift. This kind of information shows up only when data is arranged in time order.

Part four is to Aim at It: Predict the cause. With good data, the team could make a prediction about the cause of the broken belts and Roy was able to persuade Terry to check the machines on Wednesday.

Score It

Part five is to Score It: Evaluate the change. After the machines are checked by Terry's crew, Roy will be able to evaluate his prediction. We will find out the results in Chapter 3.

ACTIVITIES

Solve these problems to review what you have learned in Chapter 2. Answers are in Appendix A.

1. Find the highest, lowest, and range numbers for the following sets of data.

a. Students in last month's quality management class:
15, 25, 35, 13, 40, 12, 30, 30, 25, 20, 22, 41, 35, 16, 22, 19, 32, 27, 15, 26
_____highest   _____lowest   _____range

b. Test spins turned by sets of washing machine belt #456 before the belts began to slip: (Don't be worried because the numbers are large, the process is still simple. Look for the largest number and the smallest number and subtract to find the range.)

46,200	43,400	45,400	39,250	45,020
41,250	48,791	42,423	36,790	31,200
43,444	37,920	49,240	33,279	46,333
49,750	45,320	47,800	44,230	42,500

_____highest   _____lowest   _____range

c. Absences from work caused by the flu in January:
7, 2, 8, 5, 6, 4, 6, 5, 8, 6, 7, 4, 9, 7, 6, 7, 8, 6, 5, 4
_____highest   _____lowest   _____range

d. The number of line worker days without an accident from 1970 to 1989:

Year	Number	Year	Number
1970	40,400	1980	40,352
1971	40,250	1981	40,263
1972	40,375	1982	40,192
1973	40,450	1983	39,950
1974	40,410	1984	38,900
1975	40,350	1985	40,422
1976	40,195	1986	40,353
1977	40,475	1987	40,425
1978	40,450	1988	40,123
1979	40,332	1989	40,332

_____highest   _____lowest   _____range

2. The fire inspector is coming to do the yearly inspection of the Best Belt Company. Terry wants the safety crew to make sure everything is ready. Brainstorm a list of things the crew should check. (Work with a partner, if possible.) Review the rules for brainstorming in Chapter 1.

3. The Customer Service Department at Best Belt employs five people to answer customers' calls. Customers are complaining that they cannot get a representative when

they call. Peter sometimes hears the phones ring busily. Other times, the representatives are not on the phone at all. Peter wants to get some ideas before he meets with the representatives. The Quality Team helps him brainstorm this list:

> staff out sick
> calls about new products increase the work load
> staff is new—not trained
> broken phone needs to be fixed
> customers all call at once
> an answering machine is needed
> staff make personal phone calls
> calls about hard questions take longer
> customers like to chat
> more phone lines are needed
> staff needs more people
> calls for other departments are sent to Customer Service

   a. Do step three of evaluating the problem—put these ideas into groups. How many groups do you see? (Hint: Groups can be kinds of people, things, needs, or problems.)

   b. Step four of evaluating the problem is choosing the population Peter needs information about. List the populations (one or more) you think he should work on first.

   c. One population Peter decides to look at is kinds of calls. He wants to know how many calls are:
      1) New product related.
      2) Longer than 3 minutes (difficult questions).
      3) Sent to Customer Service by mistake.

      Here is the information the representatives gave him at the end of the week:

1) New product related calls:
   B shift: Monday 3, Wednesday 5, Tuesday 4, Friday 1, Thursday 6
   A shift: Tuesday 5, Wednesday 7, Monday 3, Thursday 6, Friday 2
   C shift: Friday 1, Thursday 3, Wednesday 2, Monday 0, Tuesday 1

2) Calls longer than 3 minutes:
   C shift: Tuesday 4, Thursday 8, Friday 2, Monday 3, Wednesday 6

A shift: Wednesday 10, Friday 2, Monday 6, Tuesday 8, Thursday 6

B shift: Thursday 10, Friday 1, Wednesday 8, Monday 5, Tuesday 6

3) Calls sent to Customer Service by mistake:

A shift: Monday 1, Wednesday 3, Friday 3, Tuesday 3, Thursday 4

B shift: Wednesday 12, Friday 10, Thursday 12, Monday 6, Tuesday 10

C shift: Tuesday 0, Thursday 1, Friday 0, Monday 0, Wednesday 1

Peter asks you to put the information in time order and make a chart for each type of call.

   d. Analyze your data to answer these questions.
      1) Which two days get the largest number of these three kinds of calls?
      2) What is the total number of calls reported to Peter?
      3) Which shift is the busiest? Which is the quietest?
   e. At Best Belt all the phone calls go through a central switchboard. Two operators are on duty on each shift. Using the data in c3, predict two causes for the B shift mistakes.
   f. What would you do to evaluate your predictions?

4) Find a process in your work setting that is often repeated. For example, you could use:

   A part that is made.

   A certain kind of phone call.

   How long it takes to deliver an order.

   a. Decide what information you want to gather about the process and write it down.
   b. Record the information for a time period (minutes, hours, days, etc.). You should have at least 20 samples.
   c. Arrange your data in a frequency distribution table. Be sure to label the rows and columns.
   d. Find the range.
   e. Can you see any significant information? Can you make any predictions from your data?

No Easy Answer

■■■■■ **GOALS**
(1) To understand the difference between coincidence and cause and effect.
(2) To use some measurable data to construct a tally chart.

WORDS AT WORK

Watch for these special words: **statistics, coincidence, just-in-time, random, tally, cell, boundary, interval, average, estimate.**

EVALUATING THE CHANGE

Score It!

George, the plant manager, has been helping Roy and Peter with the problem of broken belts. Roy reports that Terry Ndele's crew adjusted the machines on Wednesday. Then Roy kept track of the data for another week. Roy completed part five in the problem-solving process: Score It. He evaluated the change in machine setting.

"I guess we have not found the trouble," Roy says. "Peter, this week looks just like last week. There are still more broken belts on Thursday and Friday."

"Ouch!" says Peter. "What do we do now? I guess we should go back to part two: Question It."

"Let's check the material used," suggests George.

"Well, I did ask Maria about that," Roy responds. "She said

Purchasing uses two suppliers. Let's talk to her again." Peter calls Maria and asks her to join them for a short meeting.

QUESTION IT AGAIN

Cause and Effect or Coincidence?

It is very important to know about relationships in **statistics**. The data in Chapter 2 clearly showed that more belts broke on Thursday and Friday. The team knew that the machines were adjusted on the weekend. The more time went by, the more belts broke. This is a time relationship, but was the relationship a coincidence or was it cause and effect? When the machines were adjusted on Wednesday, there was no change. The belts still broke at the end of the week. It was a **coincidence**. That's something that just happens at about the same time as something else.

WORDS AT WORK

> **statistics** (sta **tis** tiks) Numbers found by looking at a small sample from a whole population. These numbers are then used to describe the larger group.
>
> **coincidence** (co **in** si dense) Two things happening at the same time or in the same place by chance. For example, two people born in the same town have a place relationship. Two people with the same birthday have a time relationship. They did not do that on purpose, or cause the other person to live in that town or be born on that day. Their relationship is coincidental.

It is tempting to look at things that have time relationships as if one caused the other. That can be dangerous. It can keep the team from looking for other causes. They could have a wrong answer that looks right. That is why Roy looked at new data after the adjustment. The new data was gathered in exactly the same way as the first data. Only the time of adjustment was different. He evaluated the change.

When Roy calls, Maria checks her computer. She brings her inventory sheets to the meeting. The sheets show that Richardson Corp. delivers fabric every Sunday night and Famous Fabric, Inc., delivers every Wednesday morning.

"That way we get our fabric **just in time** for the line to use," Maria tells the group. "It is expensive to keep a lot of fabric here."

WORDS AT WORK

> **just-in-time (or JIT)** When a company tries to have on site (in their warehouse) only as much as is needed to keep the process going, they have a just-in-time inventory.

"Would Richardson's stuff be used on Wednesday?" asks Peter.

"It might. Famous Fabric's material could hit the line at noon on Wednesday," Roy answers. "It depends on the run. That could be our problem."

"Yes," George adds, "there could be a difference in thickness. The fabric might be too thin or too thick. That could look as if the machines were to blame."

"Yeah, this time we need to look at belts that don't break. We should be sure the belt is the best thickness," says Roy. "OK, team, let's go do it!"

Choosing a Sample

"We really can't check all the belts. It will take too long. We need a sample," George tells the team.

"Yes, and it should be as **random** as possible," Maria says. "We should take a few belts from every shift and every day. We want to choose by chance (luck) within each group."

WORDS AT WORK

random (ran dum) By chance; scattered all through the population with no pattern or system. There are mathematical rules for gathering random samples in statistics.

The team asks each shift leader to take any 10 belts from the test bin for their shift each day (30 belts). George will take 10 more belts from the bin in the packing area. That will give them 40 belts each day. Over a week, they will have a sample of 200 belts to look at. The sample will not be random in a truly mathematical sense, but it will be scattered enough for the Quality Team to answer its question.

Practice: Use a deck of playing cards or a bag of multicolored covered chocolate candies to do the following:

1. Shuffle the deck and deal out four even piles, face down. Guess how many red cards will be in each pile before you look. Could you have known?
2. Before you open the bag of candies, guess how many yellow and red ones you will find. Did you guess right? Is there any way you could have known?

If you opened 1,000 bags of multicolored covered candies and counted the yellow ones in each bag, you would have a pretty good random sample of the number of yellow candies in

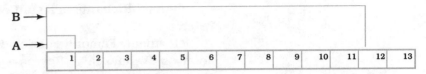

FIGURE 3.1
One millimeter compared to the thickness of the engineer's design of the belt

any bag of multicolored covered candies. That would take a lot of time. Could the Quality Team test all the belts made for a month?

Gathering the Data

The belts for the floor polisher are measured in millimeters. They were designed by the engineer to be 11.25 millimeters thick. See Figure 3.1. (If you want to know more about metric measurement, see Appendix B.)

Every day, the team carefully measured 40 belts. They used a caliper (a special kind of ruler). They rounded the measurements to one-half millimeter. They wrote down each measurement in columns. They used one column for each shift and one for the belts from the packaging bin. At the end of the week, they had 200 measurements to work with.

In Chapter 2, you learned about countable data. The thickness of the belts is a measure. This kind of data is called measurable data. The team wants to know how many belts are each measure of thickness.

SETTING UP A TALLY CHART

The team will set up a frequency distribution in the form of a **tally** chart. You may want to review Frequency Distributions in Chapter 2 before you begin this section to be sure you understand the terms. To make it easy to read the chart, the samples are divided into groups called cells. Some distributions have thousands of pieces of data in them. You can see that listing even the team's 200 pieces of data would make the chart impossible to read easily. One way to clear things up is to organize data into groups. (For example, if you had a large pile of pennies, you could count them by fives or tens—in groups.) The process of organizing data will be clearer if it is related to something you already know about.

WORDS AT WORK

tally (**tah** lee) A way of counting by making a mark for each person, thing, or event. Often a special tally sheet is used. It may have cells already on it.

Think about the last time you tried to organize your videotapes. The names of the tapes are easiest to read when they lie flat or stand up straight, in neat rows. Try this practice exercise.

Practice: (You can use a roll of masking tape to "draw" your own case on the floor or wall. Then use a ruler and some chalk to mark the shelves.)

You have a videotape case wide enough for one tape to lie flat (about 8 inches). The tapes lie on their sides so you can read the names. Each tape is an inch thick.

Your tape case is 20 inches tall. The top and bottom of the case are one-half inch thick, so there are a total of 19 inches of interior space. There are two movable shelves, each one-half inch thick. If you use both shelves, there are 18 inches of interior space. But the case has holes for the shelf brackets every half inch along the sides. There are 36 holes. See Figure 3.2.

You want to put 18 videotapes in your case. You also want to use both shelves. Aha! you think. I have 18 inches of videotapes, so they will just fit. In which hole will you set the

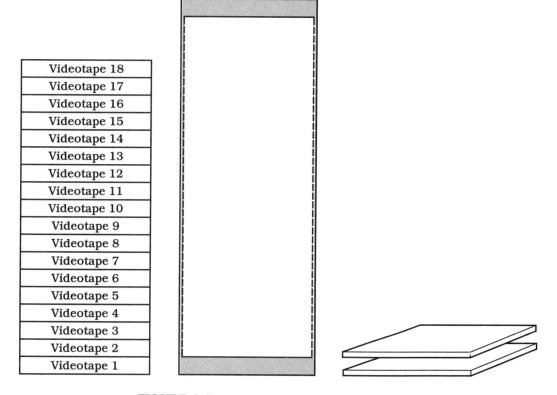

FIGURE 3.2
Your 18 videotapes, your videotape case, and 2 shelves

brackets for the first shelf? The second shelf? Is there more than one way to set the shelves? Why or why not? How many ways can you think of?

Now suppose you have three kinds of tapes. You have six adventure movies, six comedies, and six children's tapes. You want to put all of each kind together to make it easier to find what you want. Does putting each kind in a group limit how you can set up your shelves? Why?

What happens if you set your shelves in hole number 13 (6 inches from the bottom) and hole number 26 (6 inches above 13)? Can you get your 18 tapes in and out? No, you can't. There isn't enough room between the second shelf and the top of the case for the last six videotapes.

Cell Size and Boundaries

When the team sets up the tally sheet, they will group data as you grouped tapes. Instead of using types like adventure films, they will use sizes. For example, all the belts measuring 7.5, 8.0, and 8.5 millimeters could be put on the same "shelf."

They will organize the data into **cells**. Think of a cell as the space between shelves in your case. The space will be filled with pieces of data (think "tapes"). The team will also decide on how big that space will be. They will set a **boundary** (think "shelf") between each cell. The boundary cannot be exactly the same as the top or bottom number in the cell. If it were, the fit would be too "tight."

WORDS AT WORK

> **cell** (sel) A group of data organized into a unit.
> **boundary** (**bown** dah ree) Limits to the size of a cell; its top and bottom.

Suppose you want to put all the data between 5 and 10 in one cell. If you name the top boundary 5 and the bottom boundary 10, you might think that any number between 5 and 10 would fit into the cell. As you have learned, however, the boundary numbers, 5 and 10, will not fit into the cell (see Figure 3.3a). You know that from your work on the videotape case. So, you have to find a different boundary set. If you call the boundaries 4 and 11, all the numbers will fit comfortably (see Figure 3.3b).

The team has 200 observations (individuals in the sample). They need about 13 cells. There are rules in statistics for deciding how many cells to use. If you need to know how many to use, you can look up the number. All you have to know is how many pieces of data you have.

"Top Shelf" 5
Videotape 6
Videotape 7
Videotape 8
Videotape 9
"Bottom Shelf" 10

A

"Top Shelf" 4
Videotape 5
Videotape 6
Videotape 7
Videotape 8
Videotape 9
Videotape 10
"Bottom Shelf" 11

B

FIGURE 3.3
(a) One cell with boundaries 5 and 10 (b) One cell with boundaries 4 and 11

In the videotape case example, you thought of the holding place for the tapes as space. In statistics, the "space" between the boundaries is called an **interval**. Intervals may be any size that is needed to "hold" the data.

WORDS AT WORK

> **interval** (**in** turr val) The amount of space, measured in numbers, between the sides (boundaries) of a cell. All the cells have the same size interval.

When you divided the videotapes by type, there were three equal groups. The spaces between the shelves were of equal size. In statistics, intervals are required to be the same size. Think about a ruler. Each inch is the same size. There is the same amount of space between 1 and 2 inches as there is between 4 and 5 inches. What would happen if the spaces were different? The number of elements in each cell may be different, but the size (interval) of the cells must be the same.

Maria says, "Our measurements have decimals in them. It is easiest to set the interval at one whole number. That means the first cell side could be 0.75 and the second side, 1.75."

Any belt that measures more than 0.75 but less than 1.75 millimeters would be in the first cell. An interval of 1 means that one whole number will fit in that box (1.75 − .75 = 1). All the belt measurements are rounded to the nearest one-half millimeter. They will be some number followed by .0 or .5. (If you need to know more about decimals, see Appendix B.)

Setting Up Cells "First," Roy says, "we need to find the thinnest and the thickest measures in our data."

"The thinnest I can find is 5.0," says Peter.

"I've got the thickest, 17," Roy adds. "So our cells have to go from 4.75 to 17.75."

"That's right. And we need a midpoint for each cell to mark the tally on," Maria adds. "That way, the data will look much clearer."

When numbers are tallied at the midpoint, they are easier to count. It's as if you took all the numbers in each cell and **averaged** them. If you do that, they will come out very close to the midpoint of the cell.

WORDS AT WORK

> **average** (a ver age) A middle point in a group of numbers. Averages are found by adding all of the sample numbers together and then dividing that total by the number of samples.

If the team sets up the cells to have boundaries at .75 of every millimeter, the midpoint will be halfway between 4.75 and 5.75 for the first cell. Let's divide the cell into equal parts, beginning with 4.75, as shown in Figure 3.4. What will be the midpoint? That's right, 5.25 [(4.75 + 5.75) ÷ 2]). The team can mark all the measures in this cell at the midpoint, 5.25.

It is important to understand that statistics are **estimates.** Statistics describe the population as truly as possible without inspecting every individual in it.

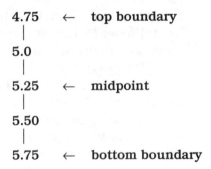

FIGURE 3.4
Cell divided into equal parts, beginning with 4.75 and ending with 5.75. The midpoint is 5.25.

NO EASY ANSWER

WORDS AT WORK

estimate (s ti mate)   A number that is close enough to give good information but is not completely accurate.

Averages and midpoints bring us close to the exact measures. There is always some small difference between the samples as tallied and the actual population. There are many ways to correct for most of the differences.

"Now, we're ready to set up our cells and begin to tally," says George. "Here's what the tally sheet looks like." (See Figure 3.5.)

TALLY OF BELT THICKNESS

		Monday	Tuesday	Wednesday	Thursday	Friday	Total
Cells							
	4.75						
#1	5.25						
	5.75						
#2	6.25						
	6.75						
#3	7.25						
	7.75						
#4	8.25						
	8.75						
#5	9.25						
	9.75						
#6	10.25						
	10.75						
#7	11.25						
	11.75						
#8	12.25						
	12.75						
#9	13.25						
	13.75						
#10	14.25						
	14.75						
#11	15.25						
	15.75						
#12	16.25						
	16.75						
#13	17.25						
	17.75						

FIGURE 3.5
The Quality Team tally sheet

NO EASY ANSWER

12.0	15.0	11.0	11.0
13.0	15.5	17.0	9.5
15.5	10.0	15.0	13.0
17.0	12.0	12.5	10.0
14.0	8.5	8.0	14.0
13.5	12.5	13.5	11.5
9.0	13.5	15.5	11.5
12.0	12.5	9.5	13.0
12.5	12.5	13.5	9.0
12.5	11.0	11.5	14.5

FIGURE 3.6
Monday's data

TALLY OF BELT THICKNESS

Cells		Monday	Tuesday	Wednesday	Thursday	Friday	Total
	4.75						
#1	5.25						
	5.75						
#2	6.25						
	6.75						
#3	7.25						
	7.75						
#4	8.25	— //					
	8.75						
#5	9.25	—////					
	9.75						
#6	10.25	— //					
	10.75						
#7	11.25	—𝐽𝐻𝐼 /					
	11.75						
#8	12.25	—𝐽𝐻𝐼 ////					
	12.75						
#9	13.25	—𝐽𝐻𝐼 //					
	13.75						
#10	14.25	—///					
	14.75						
#11	15.25	—𝐽𝐻𝐼					
	15.75						
#12	16.25						
	16.75						
#13	17.25	—//					
	17.75						

FIGURE 3.7
Monday's tally on the sheet

NO EASY ANSWER

"OK, Roy, let's have the list for Monday. I'll tally by placing a mark for each observation and a slash each time we get to five in the same cell," Peter volunteers. See Figures 3.6 and 3.7 for the results.

SUMMARY

The team has learned about a time relationship—when two things happen at once or follow one another. They found out that adjusting the machines did not make the belts last longer. So the problem was not solved. When two things just happen together, that is called a coincidence. It is not wise to think coincidences are cause-and-effect events without more proof.

The team then decided to check the result of using materials from two suppliers. The suppliers brought belt fabric just in time as production needed it. Two hundred samples were chosen at random from the week's work. Each was measured in millimeters. The team set up a tally chart. They set cell boundaries and midpoints. They chose a cell interval (size) of one. Peter tallied Monday's observations.

In the next chapter, the team will use the tally results to find out if their belts are the right thickness. They will look at other measures of the middle. They hope to find the answer to their problem soon.

ACTIVITIES

See Appendix A for answers and discussion.

1. Here are Tuesday's observations. Using Figure 3.7, put them on the tally sheet. (Copy it if you can, so you will have extras.)

13.0	13.5	15.0	13.0
13.0	10.5	13.0	13.5
9.5	12.0	7.5	11.0
9.5	10.0	12.0	14.5
14.5	12.5	12.0	12.0
14.0	14.0	13.0	15.0
15.0	11.0	11.5	10.0
11.5	14.0	14.5	10.0
13.5	8.0	13.5	9.0
11.5	11.5	12.5	14.5

  a. What are the lowest and highest of Tuesday's observations? What is the range?
  b. Find the average thickness of Tuesday's belts. (Add the samples together with your calculator, or by col-

umn. Then, when you have added them all together, divide that number by the number of belts in Tuesday's sample. (If you are not sure where to put the decimal point, see Appendix B.)

2. Finish the following story by putting the right words in the blanks. Use the list below.

   Connie wants to find out if her advertising is bringing new customers to Best Belt. She decides to analyze the phone calls coming in to the sales office. The total number of phone calls is the _____. Each telephone call is an _____ in the sample. The new ad is in the trade magazine on Monday. Connie asks her staff to make a record of all the calls they receive the rest of the week. She wants them to divide the calls they get into three groups:

   a. Regular customers who saw the ad.
   b. New customers who saw the ad.
   c. New customers who did not see the ad.

   She asks the staff to keep a _____ for four days: Tuesday, Wednesday, Thursday, and Friday. Connie believes the _____ about the sample will give her good information about the whole population. She wants to be sure that the relationship between new customers and the ad is not just a _____.
   The staff makes a tally sheet with six _____. The cell _____ are 0, 5, 10, 15, 20, and 25. That means that the cell _____ is 5. To make the data clear, the numbers are tallied at the cell _____ halfway between each boundary. Connie's staff finds the midpoints by getting the _____ of two cell boundaries.

> Word List: midpoint, interval, tally, cells, boundaries, population, statistics, coincidence, average, individual.

3. Find three time relationships in your workplace. Do they show cause and effect or are they coincidence? Why? (Hint: A coincidence might be "Every time I start to talk to someone, the phone rings.")

4. Use the data you collected about your workplace for Activity no. 4 in Chapter 2. Set up a tally sheet with cells. Choose a cell interval and cell boundaries. Find the cell midpoint for each cell. How many cells did you decide on?

5. Choose another set of data from your workplace. Pick one that will show a "problem." To do this activity, you will have to follow the problem-solving steps. Fill in the information below to show how you might go about solving your problem.
   a. Part 1   See It: Evaluate the problem.
   Write the problem. (Remember the four steps: identify the problem, get ideas (brainstorm), group them, choose the population.)
   b. Part 2   Question It: Gathering good information.
   Write three questions you could ask about the problem. How will you get the answers? List the steps you will take to gather data. If possible, gather real data about the most important question. If you cannot get real data, make up at least 30 pieces of information about a population to create a sample.
   c. Part 3   Think about It: Analyze the data.
   Arrange your data in a frequency distribution in the form of a tally sheet. What information does your distribution give you? Does it show a significant relationship between two things? Why or why not? What data would show that the relationship is not a coincidence?
   d. Part 4   Aim at It: Predict the cause.
   What does your data tell you might be the cause? If it is not clear, what do you think the cause is? Write your prediction. What change will you make in an effort to solve the problem?
   e. Part 5   Score It: Evaluate the change. What happened when you made the change? Did the change solve the problem? Suppose the change did not work. What would you do next?

4

Measures of the Middle

■■■■■ **GOALS**
(1) **To understand measures of the middle: mean, median, and mode.**
(2) **To recognize a bell curve (normal distribution).**

WORDS AT WORK

tolerance, normal, central tendency, mean, median, mode, outlier, bell curve

After Monday's and Tuesday's samples are tallied, the Quality Team has a question. Are these belts the right thickness? George says, "The engineers designed the belts to be $11\frac{1}{4}$ millimeters thick." You helped tally, so you know the belts are not all 11.25 millimeters thick. Monday's belts have a 9-millimeter range (17.0 − 8.0 = 9.0) How close are the belts to the expected thickness? How close do they have to be? Look at the tally sheet for Monday (Figure 3.7). How many belts were between 10.75 and 11.75? Right, six.

TOLERANCES   In Chapter 1, you learned about specifications. The belt's specification is 11.25 millimeters thick. You learned that tiny differences always exist among belts. No two can be exactly the same. Some differences are significant (important). Some are not. When the thickness is "in control," the differences are not significant. How can the team know? They need to find out from the engineers.

The engineers know how much the thickness can vary and still be within **tolerances.** If you tolerate something, you may not like it, but you can live with it.

WORDS AT WORK

> **tolerances** (**tol** er n ses)   The acceptable limits; not exact but close enough. In every sample, some individuals are "okay." They are not too large or too small to work well. Sometimes tolerance and specification are used to mean the same thing. They both mean within usable limits.

Within the tolerances, the belts will still work fine. When the belts are outside the upper or lower limit, they may not work well. The goal of total quality control is to have all the belts work well. The engineers at Best Belt ran tests to set the tolerances. The belts were good when they were between 9 and 14 millimeters thick. You could say the range of thickness of good belts is 5 millimeters. See Figure 4.1.

When you want to know if a large sample of belts is good, there are several things you could look at. You could look at the average thickness. You learned about averages in Chapter 3. Another word for average is *mean*. In Chapter 3 you figured the mean of Tuesday's belt sample to be 12.25. This is within tolerances. Does this mean that the sample as a whole is good?

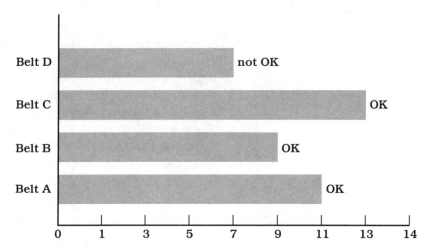

FIGURE 4.1
Belts and tolerances

MEASURES OF THE MIDDLE

WORDS AT WORK

mean (mee n) Average.

How many of Tuesday's belts are 11.25 millimeters thick? According to Figure 4.2, six belts were tallied at 11.25. How many belts are within the tolerances? Right, 3 + 4 + 6 + 9 = 28. Twenty-eight out of forty belts are in control. That is 70 percent. It looks as if Best Belt has a way to go before the pol-

TALLY OF BELT THICKNESS

Cells		Monday	Tuesday	Wednesday	Thursday	Friday	Total
	4.75						
#1	5.25						
	5.75						
#2	6.25						
	6.75						
#3	7.25		/				
	7.75						
#4	8.25	//	/				
	8.75						
#5	9.25	////	///				
	9.75						
#6	10.25	//	////				
	10.75						
#7	11.25	ꟾꟾꟾꟾ /	ꟾꟾꟾꟾ /				
	11.75						
#8	12.25	ꟾꟾꟾꟾ ////	ꟾꟾꟾꟾ /				
	12.75						
#9	13.25	ꟾꟾꟾꟾ //	ꟾꟾꟾꟾ ////				
	13.75						
#10	14.25	///	ꟾꟾꟾꟾ //				
	14.75						
#11	15.25	ꟾꟾꟾꟾ	///				
	15.75						
#12	16.25						
	16.75						
#13	17.25	//					
	17.75						

FIGURE 4.2
Tally for Monday and Tuesday

isher belts are in control. How many of Tuesday's belts are too thin? How many are too thick? Is it safe to think all the belts are good if the average is within tolerance? The team needs to know how far away from the engineers' mean (average) the belts really go. They also need to know if the belts are above or below the mean.

If the belts are too thin, they will break too soon. If they are too thick, they could cause a different problem. Peter does not want any more problems!

CENTRAL TENDENCY

The team decides to look at two other ways to consider the middle measurement of the sample. In everyday life, we talk about things being **"normal."** What do we mean by that?

WORDS AT WORK

> **normal** (nor mal)   The way things happen most of the time. In a normal distribution, the number with the highest frequency is close to average.

In any very large number of things or people, most will be closer to the average than to the ends of the distribution. For example, most people born in the United States are between five and six feet tall. Suppose you checked the height of everyone at a Super Bowl game. The average height would probably be around $5\frac{1}{2}$ feet. A few people would be much taller, and a few would be much shorter. This grouping of people and things around the middle is called the **central tendency.**

In the total population of Best Belt #295, most of the belts should be around average; that is, 11.25 millimeters thick. That is the engineers' "perfect" belt.

WORDS AT WORK

> **central tendency** (sen trall ten den c)   When a large number of measures is taken, most will be close to the middle of the range. A few will be at each end of the range.
> **median** (mee dee n)   The middle value in a row of numbers. Half the measurements are less than the median and half are greater.
> **mode** (mo de)   The number that shows up most often in a sample. It does not have to be near the median, and it may move the average away from the median.

MEASURES OF THE MIDDLE

FIGURE 4.3
Distribution of heights at the Super Bowl (© John W. Mayo/UNICORN STOCK PHOTOS)

Two other ways to measure the central tendency are the **median** and the **mode.** Think about the highway median. It is in the middle of two equal sides of a highway. Often there are two lanes on each side.

The Median The median divides the belt thicknesses in half. Just as many are thicker than the median as are thinner than the median. Put an odd number of individuals in numerical order. The median will be the middle number. For instance, list nine numbers in numerical order. The median is whatever the fifth number is. The numbers might be 3, 4, 4, 5, 6, 7, 8, 8, and 9. The median is 6. Four numbers are less than 6, and four are greater. Figure the mean of this set of data. The total, 54, divided by 9 equals 6. In this set the mean and median are the same. That is not always true.

Look at this set: 2, 3, 3, 4, 6, 9, 10, 12, 14. Find the median. Right, it is 6 again. Figure the average. That's right, the average (mean) is 7. The number 7 is not even in the set. The median shows the middle measurement, not the average. The median can be one of several identical numbers. It can also be between two numbers.

What do you do if a set contains an even number of measurements? Look at this set: 2, 3, 3, 4, 5, 7, 8, 8, 9, 9. Here there are four numbers less than 5. There are four numbers greater

■ 41 ■

than 7. Which is the median? The median, in this case, is the average or mean of the two center numbers, 5 and 7, which is 6. Like the mean, the median does not always show up in the set.

The median is useful when a few very large or very small numbers make the set mean off center. Look at this set: 2, 3, 3, 4, 5, 6, 6, 7, 30. Here the median, 5, is a good measure of the central tendency. Figure the mean. (The total, 72, divided by 9 numbers in the set, equals 8.) All but one of the numbers in the set is less than the mean. The mean is not a good measure of the central tendency.

In statistics, a strange number like the 30 is often called an **outlier**. If there is only one outlier, it might not be significant. In real life, the team would want to be sure the outlier happened by chance.

WORDS AT WORK

outlier (**out** lie er) A single number that is much farther from the mean than any other number in the set. It may be at either end of the distribution.

The Mode

Another way of looking at the central tendency of a distribution is to look at the mode. The mode tells which number in the set happens most often. In the distribution 3, 3, 4, 4, 4, 6, 6, 6, 6, 6, 7, 7, 7, 9, the mode is 6. It is the number that shows up most frequently. Find the mode in this set: 7, 9, 6, 3, 6, 4, 7, 6, 3, 6, 4, 7, 4, 6. Right, it is 6. (It helps if you put the set in order.)

Sometimes the mode is near the center. Sometimes it is not. In this set, 3, 3, 3, 3, 3, 3, 6, 6, 7, 10, 10, 16, 18, the mode is 3. The median is 6, and the mean is 7. Here, the mode is not a good measure of the center. It just says that 3 happens more often than any other number. That may be important. Three may be what you want. If you want all the numbers to be 3, you have a problem. More than half the set is very different from 3. Suppose you were making belts 3 millimeters thick. The sample you took looked like this set of numbers. Do you think the machine might need to be fixed?

PLOTTING A NORMAL DISTRIBUTION

When the engineers designed the polisher belt, they set the tolerances. Then they made a normal frequency distribution of a large number of belts. Figure 4.4 shows what a sample of 100 normally distributed belts looks like. In this sample, the mean, the median, and the mode are all 11.25.

MEASURES OF THE MIDDLE

TALLY OF BELT THICKNESS

Cells:			Total:
	7.75		
#1	8.25	/	1
	8.75		
#2	9.25	|||| |||| ||||	15
	9.75		
#3	10.25	|||| |||| |||| //	17
	10.75		
#4	11.25	|||| |||| |||| |||| |||| |||| ////	34
	11.75		
#5	12.25	|||| |||| |||| //	17
	12.75		
#6	13.25	|||| |||| ||||	15
	13.75		
#7	14.25	/	1
	14.75		100

FIGURE 4.4
The engineers' sample

To find the mean, first total all the measurements, as follows:

$$
\begin{aligned}
8.25 \times 1 &= 8.25 \\
9.25 \times 15 &= 138.75 \\
10.25 \times 17 &= 174.25 \\
11.25 \times 34 &= 382.50 \\
12.25 \times 17 &= 208.25 \\
13.25 \times 15 &= 198.75 \\
14.25 \times 1 &= \underline{14.25} \\
&\ 1125.00
\end{aligned}
$$

Then divide the total by the number of measurements, 100 (1125 ÷ 100 = 11.25).

To find the median: Determine which measurement(s) fall in the middle. (There are 49 measures less than the 50th, and 49 measures greater than the 51st.) The 50th and the 51st are both 11.25 so there is no need to get an average. The median is 11.25. Find the mode: The most frequent measurement is 11.25.

A normal distribution can be shown on a graph, as shown in Figure 4.5. If turned on its side, the frequency distribution makes the shape of a bell. Normal distributions are of-

MEASURES OF THE MIDDLE

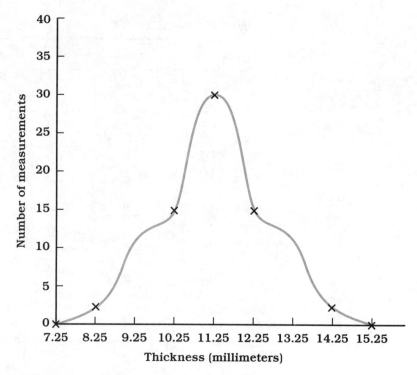

FIGURE 4.5
A bell curve showing the normal distribution of good polisher belts.

ten called **bell curves.** If a process is in control, its graph should look like a bell. It may not be shaped exactly like this one. Some bells are tall and thin. Some are short and fat. The shape depends on how big the range of the sample is. In the next chapter, the team will compare their data to the engineers' bell curve.

WORDS AT WORK

bell curve (bell kur ve) A normal curve that is shaped like a bell; both sides are equal.

SUMMARY In this chapter, the team checked the tolerances of the belt specifications. They compared their sample to the engineers'. Now they know that many of the polisher belts are not within tolerance. They vary too much.

The team looked at the measures of central tendency. They know the mean is the average, the median divides the samples in half, and the mode tells which measurement is most fre-

MEASURES OF THE MIDDLE

quent. The median and the mode are useful when a few measurements, called outliers, throw the mean off center. In a normal or bell curve, the mean, median, and mode are the same.

ACTIVITIES

See Appendix A for answers and discussion.

1. You decide to check three shifts for two days. You find these belt thicknesses:
   Monday: Shift A:13, 10, 13, 11, 13; Shift B: 11, 14, 13, 15, 11; Shift C: 13, 15, 11, 14, 13
   Tuesday: Shift A: 12, 14, 16, 13, 12; Shift B: 15, 14, 13, 15, 12; Shift C: 14, 13, 12, 13, 12
   Use the data for both days:
   a. What is the range?
   b. What is the mean?
   c. Arrange the data in numerical order. What is the median?
   d. What is the mode?

2. Using the data in item 1, make a tally sheet. (You will not need cells for a small sample like this.) Turn the frequency distribution on its side. Draw a curve line above the tally. Is this a normal distribution? Why or why not?

3. a. Use the data below. Find the mean, median, and mode.

   13, 12, 13, 12, 13, 12, 13, 13, 12, 13, 13, 12, 13, 13, 14, 13, 13, 14, 13, 13, 14, 13, 13, 14, 13, 13, 14, 13, 13, 13

   mean _____ mode _____ median _____

   b. Make a tally sheet. Turn it on its side. Draw the curve. Is this a normal distribution?

4. a. Use the data below. Find the mean, median, and mode. Is there a clear mode? Why or why not?

Shift	Monday	Tuesday	Wednesday	Thursday	Friday
A	1	4	3	8	8
B	2	5	3	7	8
C	2	3	6	6	9

   b. Make a frequency distribution. Turn it on its side and draw the curve line. Is this a normal distribution? Why or why not?

5

Solving the Problem

▪ ▪ ▪ ▪ ▪ **GOALS**
(1) To read a line graph.
(2) To understand the measure of spread (dispersion), called a standard deviation from the mean.

WORDS AT WORK

axis, horizontal axis, vertical axis, dispersion, standard deviation, probability

READING A LINE GRAPH

In Chapter 4 you turned the frequency distribution in Figure 4.4 on its side and created the line graph in Figure 4.5. A line graph shows information as a picture. A line graph shows the frequency tally without the tally marks.

    Line graphs have a special shape. There are two lines that meet in a corner, forming an L shape. These lines are called **axes,** a horizontal axis and a vertical axis. Each axis has a job to do. In most frequency distribution line graphs, the vertical axis shows the number of individuals in a sample. On the engineer's and the team's graphs, the vertical axis shows the number of belt measurements counted by fives. The horizontal axis usually shows some kind of measure. On these graphs, the horizontal axis shows the thickness of the belts in

▪ 47 ▪

millimeters. The horizontal axis can show measures like time (number of days), or distance (number of inches or miles).

An axis can show anything that can be written in a standard measure. The team could have put the thickness on the vertical axis and the number of belts on the horizontal axis. It is the relationship between the two axes that counts. Here, the team is relating the number of belts to their thickness. Later, you will use line graphs to show other relationships.

WORDS AT WORK

> **axis (ax** iss) Each of the two lines in a line graph. (Note that the plural of axis is axes.) The lines are either vertical or horizontal.
> **vertical (ver** ti cal) A line that runs up and down.
> **horizontal (hor** i zon tal) A line that runs from left to right, as does the "line" where the earth meets the sky—the horizon. So any line like that is called horizontal.

To read a line graph, choose any point on the graph line and read horizontally and vertically to the corresponding measures on the two axes. (Hint: It helps to use a ruler to form a straight line.) For instance, at the highest point in Figure 5.1, you can see that with a thickness of 11.25 millimeters (reading down to the horizontal axis), there are 34 belts (reading across to the vertical axis) in the engineers' sample of 100 belts.

Setting points on the axis is like setting cell size and

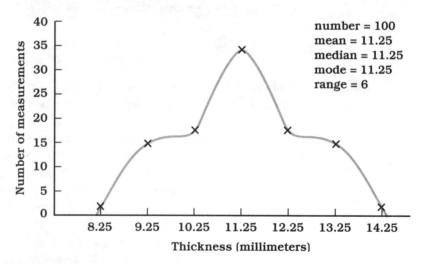

FIGURE 5.1
Engineers' sample of 100 as a curve

boundaries. The points on the axis can be any number you choose. However the intervals must be the same size (standard). The points must all have the same value between them. On these graphs, there are five belts between each point on the vertical axis. There is one millimeter between each point on the horizontal axis.

The team have decided to compare their data to the engineers' normal distribution. In a perfect normal distribution almost all the belts will be between 9 and 14 millimeters thick. Let's see what they find.

Figure 5.1 shows the engineers' normal frequency distribution as a bell curve with a mean of 11.25. Notice the shape of the bell. The left side of the bell (8.25 to 11.25) is exactly the same shape as the right side (11.25 to 14.25). Make a copy of the page. If you fold the bell in half along the 11.25 line, the sides will match. This is how a normal distribution curve always looks.

Figure 5.2 shows the week of samples collected by the team. The team will use the Total column for their comparison.

Look at Figure 5.3. This shows the curve for the week of samples Roy took and Peter tallied. Notice that the shape is not a perfect bell. The side to the left of the mean has a different shape from the right. The bell has a few "dents" in it. Most of the belts are within the tolerances, 9 to 14 (143 belts), but many are not (57). Thirty-two are too thin, and twenty-five are too thick. The mean of the team's sample is 11.16.

DISPERSION OR SPREAD

The sample Roy got spreads out along the horizontal axis. Roy's data in Figure 5.3 spreads out farther (5.25 to 17.25) than the engineers' data in Figure 5.1 (8.25 to 14.25). In the engineers' sample distribution, only one belt was too thin and one belt was too thick.

Remember, the team wants to know how far from the mean their data went. They also want to know in which direction the data went. That is called the **dispersion**. Looking at Figure 5.3, the team can tell that their average is close to the engineers' mean, but their range is larger. Their range is 12 (17.25 − 5.25) compared to the engineers' range of 6 (14.25 − 8.25). The team can see that their dispersion is greater. They know that averages alone do not tell the whole story.

WORDS AT WORK

dispersion (dis **per** zyun) How far the data spreads on the horizontal axis.

TALLY OF BELT THICKNESS

		Monday	Tuesday	Wednesday	Thursday	Friday	Total
Cells							
	4.75						
#1	5.25					/	1
	5.75						
#2	6.25					/	1
	6.75						
#3	7.25		/	/	///	ЖЖ /	11
	7.75						
#4	8.25	//	/	//	ЖЖ /	ЖЖ ///	19
	8.75						
#5	9.25	////	///	ЖЖ	/	ЖЖ	18
	9.75						
#6	10.25	//	////	ЖЖ ЖЖ //	ЖЖ ЖЖ ЖЖ	ЖЖ ///	40
	10.75						
#7	11.25	ЖЖ /	ЖЖ /	ЖЖ	ЖЖ //	ЖЖ	29
	11.75						
#8	12.25	ЖЖ ////	ЖЖ /	ЖЖ ЖЖ	ЖЖ	////	33
	12.75						
#9	13.25	ЖЖ //	ЖЖ ////	/	////	//	23
	13.75						
#10	14.25	///	ЖЖ //	///			13
	14.75						
#11	15.25	ЖЖ	///	/			9
	15.75						
#12	16.25			/			1
	16.75						
#13	17.25	//					2
	17.75						

FIGURE 5.2
The week's tally of belt thickness

Figure 5.4 shows two distributions with the same number of items and the same mean. These are the distributions you worked with in the Activities section of Chapter 4. Are the distributions the same? Why or why not? Right, the spread or dispersion is different. In distribution A the 30 measurements are dispersed more widely than in distribution B.

If the tolerance for these items is between 10 and 16, both show good items only. If the tolerance is between 11 and 15, distribution A shows some bad items. If all you know are

SOLVING THE PROBLEM

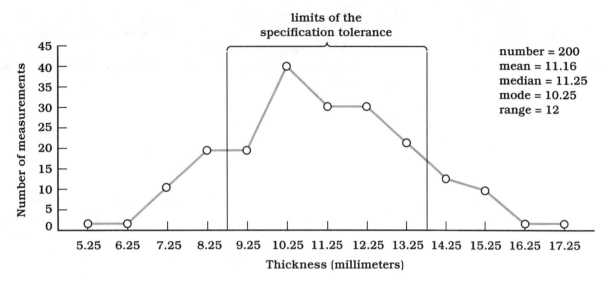

FIGURE 5.3
The team's sample total as a curve

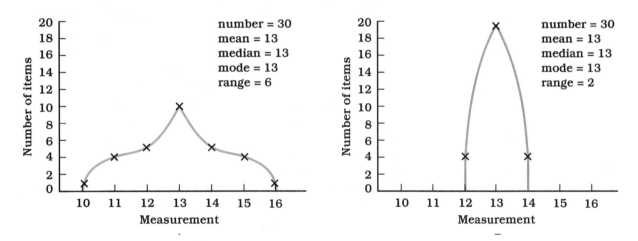

FIGURE 5.4
Distribution with like means

mean, median, and mode, you cannot tell if all the items are good. Knowing the range will help some. In A, the range is 6. In B, the range is 2. If the range of tolerance is 3, you know A has some bad items. You do not know exactly where they are. Because the mean is where it should be, at 13, you can guess that some bad items will be on each side of the mean.

Suppose the range of tolerance is 4. What does that tell you about sample B? It is very tightly controlled. A slightly greater dispersion would still contain all good items. Perhaps

SOLVING THE PROBLEM

the product could cost less to make if it were not so tightly controlled. It is best to have the sample look like a bell curve within the tolerance limits.

The team cannot tell how A is different from B if they use only the mean.

STANDARD DEVIATION

Roy and Peter know the way to compare sample curves in statistics. It is called the **standard deviation**.

WORDS AT WORK

> **standard deviation** (stand r d dee vee **a** shun) A measure of distance from the mean. It is used to understand the dispersion of a sample. It is shown by the Greek letter sigma (σ).

A deviation is any change or difference from the mean. Standard means even and regular. A ruler measures things using a standard like the foot (12 inches) or the meter (100 centimeters). Each inch is exactly the same size as every other one. Each centimeter is the same size as every other one. Inches and centimeters are standards. If inches were all different sizes, how would we measure anything?

A standard deviation measures how far individuals in a distribution are from the mean. It shows how well a process is in control. Remember, "in control" is when almost all the belts are good belts. They meet the specifications. They are within the tolerance limits.

The bell curves in Figure 5.5 show a normal distribution. Notice that all three curves are the same shape. The only difference is in the size of the shaded area. The shaded areas illustrate the **probability** of occurrence within that area.

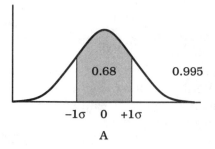
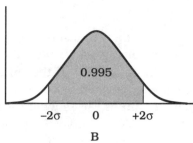
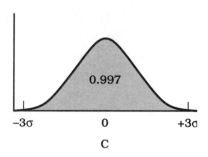

FIGURE 5.5
Standard deviation on a bell curve

PROBABILITY IN THE NORMAL DISTRIBUTION

Figure 5.5 illustrates that in a normal distribution, 68 out of every 100 (68/100 or .68 or 68 percent) of the sample items will be found between –1 and +1 on the horizontal axis. That is, it is probable that 68 percent will be between one deviation to the left and one deviation to the right of the mean (shown as 0 on the graph). It is also probable that 995 out of every 1,000 items will be found between –2 and +2 standard deviations from the mean. By the time the third standard deviation is reached, 997 out of 1,000 items will be included. Only 3 will be farther from the mean. People who work with statistics say, "The probability is that 997 items in our sample of 1,000 will fall between –3 and +3."

WORDS AT WORK

> **probability** (prah bab **il** i tee) The likelihood that something will happen. Another word commonly used is *odds*. In the United States, there is a greater probability of snow in January than in June. The odds are better that it will snow in January than in June.

How can the team find out which number to use as the standard deviation for its sample? In other words, how can they find out where to locate points +1 and –1, and +2 and –2, and +3 and –3 on the horizontal axis? There is a mathematical formula. At this point, the team knows

- All the belt thicknesses in their sample,
- What size cell to put them in,
- The midpoint of the cell to tally them on,
- The number of belts at each midpoint, and
- The mean of all the daily samples.

Now, they can use the formula to find the standard deviation, *sigma* (σ), for their sample. When you take a statistics course, you will learn how to use the algebra formula to find the standard deviation.

The engineers' sigma is 1.32. There were 100 belts in the normal distribution they created. By the rule of standard deviation for normal distributions, 68 of them (68 percent) should be between 11.25 – 1.32 (9.93) and 11.25 + 1.32 (12.57), and 99 (99.5 percent) should be between 11.25 – 2.64 (8.61) and 11.25 + 2.64 (13.89). Look at the frequency distribution in Chapter 4, Figure 4.4. Are about 68 belts thicker than 9.25 and thinner than 13.25? Yes. Are about 99 belts thicker than

8.25 and thinner than 14.25? The graph shows 98. Are any belts farther from the mean than three sigma? That would be thinner than 8.25 or thicker than 14.25. There are none in this sample because it is only 100 belts.

If there were 1,000 belts in this sample, the team would expect to find 3 (1,000 − 997) out of 1,000 outside the third sigma. It is probable that 997 of 1,000 will be within plus and minus three sigma. (If you want to know more about positive (plus) and negative (minus) numbers, see Appendix B.)

When the engineers made their curve, they based it on a daily production of 5,000 #295 belts in a stable process. A stable process is a process that does not change. It is always done the same way. It should always give the same result. If you used a population of 5,000 as the sample, there would be 3,400 belts between −1 and +1 (68 percent of 5,000). There would be 4,975 belts between −2 and +2 (99.5 percent of 5,000). Also, there would be 4,985 belts between −3 and +3 (99.7 percent of 5,000). In a normal distribution when the process is stable, the engineers expect 15 belts to be outside the third standard deviation. The shape of that curve is a perfect bell like those in Figure 5.5. Using a small sample like 100 in the engineers' sample makes the bell sharper in shape. Looking at curves and standard deviations is a way of estimating what will actually happen.

USING ESTIMATES

You may have read that the average American family has 2.3 children. You know there cannot be $\frac{3}{10}$ of a child living in a family. The number is an average or mean. Say ten families have 4, 2, 1, 5, 4, 2, 1, 2, 1, and 1 each. There are a total of 23 children. Find the average (23 ÷ 10 = 2.3). None of the families really has part of a child. The average is an estimate. So is the standard deviation. It tells about how many individuals in a sample are between points. Remember when the team grouped the tally at the cell midpoints? That made a kind of estimate. Not all the belts in the cell with boundaries of 11.75 and 12.75 are 11.25 millimeters thick. The team's purpose is to find out why the belts are breaking. They can do that without keeping every exact measurement. The cell midpoint is close enough. All of statistics is about being "close enough" to get useful information. It is not necessary, or even possible, to be absolutely accurate.

COMPARING CURVES

Figure 5.6 shows the data the team got compared to the engineers' sample (doubling the results of the engineers' 100-belt sample to make it comparable to the team's 200-belt sample).

SOLVING THE PROBLEM

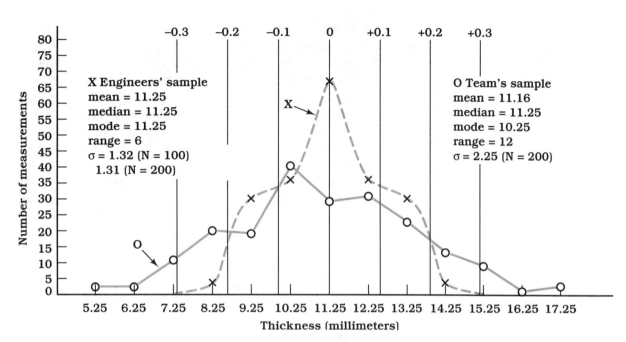

FIGURE 5.6
Engineers' sample compared to team's sample

Without seeing this graph, you know that the two curves are not exactly alike. You can see a small difference in the means (11.25 − 11.16 = .09). The range of the team's data is much larger (12, not 6). What does that tell you? Right, some of Roy's belts will be farther from the mean than the belts in the engineers' sample. So you can predict that Best Belt's production process is not as much in control as the engineers want it to be. Once the team draws the curves, they can see the difference easily.

USING THE DATA TO FIND A REASON

Do you see from Figure 5.6 why Peter is having a problem? Compare the belts made by Roy's department to the engineers' distribution. The team can see that 57 out of their 200 belts are not within the tolerances. Some are too thin and some are too thick.

Look at the frequency distribution in Figure 5.2. You will remember from Chapter 2 that more belts broke on Thursdays and Fridays than on Mondays and Tuesdays. At the bottom of each column, find the mean. Do the belts get thinner as the week goes on? Look at each day's frequency distribution. Notice that Monday and Tuesday have fewer low measurements and more high measurements than Thursday and Friday. On Thursday and Friday, there are no belts thicker than 14 millimeters (the upper tolerance) and 25 belts thinner than 9 millimeters (the lower tolerance).

The team knows that too-thick belts can cause problems. However, they are probably going to take longer to break. The thin belts are most likely to break too soon.

The team has enough information now to try part four of the problem-solving process again, predicting the cause.

At the next meeting, Maria says, "The daily frequency chart shows more thin belts at the end of the week. We're showing 25 too-thin belts out of 80 for Thursday and Friday. Let's leave out Wednesday because the fabric supplies are mixed on Wednesday. On Monday and Tuesday, we're showing only 4 too-thin belts out of 80 samples."

Peter says, "Yes. We know from the double curve line graph that the process is not in control. We also know that the problem is not caused by the machines. We checked that out."

"And it's not my people, either," Roy adds. "The first work we did showed no difference among the shifts."

"I called the customers just to be sure," Connie says. "They are all using the correct belt."

"So, what do you think?" asks George. "Is it the material?"

"It could be," says Roy. "We have ruled out everything else. Let's review once more. First, we ruled out worker mistake. All three shifts had about the same number of belts break for the three weeks we sampled. And all three shifts had more broken belts on Thursday and Friday.

"Second, we thought the machines were losing adjustment from Monday to Friday. Maintenance checked them on Wednesday. The results of the next week were the same. More belts broke on Thursday and Friday. So, it's not the machines.

"Third, we took a sample of 200 belts, 40 each day for five days. We measured each one to see how thick it was. We set up cells for grouping and chose the midpoint. Peter tallied the measurements for each day. Then we got the totals for the week.

"We found the range, mean, median, and mode of our sample. We made a line graph of the sample from the frequency distribution.

"Fourth, we compared our line graph to the engineers' graph. Their graph shows how the sample looks when it is in control. By putting both lines on the same graph, we could see the difference clearly. We saw that our belts are both thinner and thicker than they should be.

"Now we know the problem. Look at the tally sheet [Figure 5.2]. Richardson's average thickness is 12.45 on Monday and 12.25 on Tuesday. Famous Fabric's average thickness is 10.68 on Thursday and 9.52 on Friday. The averages are within tolerance, but look at the dispersion. By comparing curves, we can tell how different our production is from the engineers' design. The averages alone do not give us the whole picture."

"OK," says Maria. "I guess I have to talk to the suppliers right away. I'll talk to Famous Fabric first. Their thinner fabric seems to be causing Peter's problem. I will explain that they must make sure the fabric is at least 9 millimeters thick. Their fabric is all less than 14 millimeters thick. They have done a good job at that end. I will let them know we are pleased about that.

"I will also ask Richardson to be sure their fabric is within specs. It should never be more than 14 or less than 9 millimeters thick. I'll tell them we're concerned that too-thick fabric might cause another kind of problem."

"Let me know how both companies react," says George. "By the way, Peter has replaced the broken belts for our customers. We want them to be happy with our service. After the suppliers adjust the material we will do another sample. We want to be sure we found the problem."

Best Belt Company uses a just-in-time (JIT) inventory system. That way, there is not very much material waiting to be used by Roy's department. The fabric companies agree to adjust their processes right away. When the team meets two weeks later, Roy shows them his new statistics.

"As you can see from the comparison chart," he says, "there is now no real difference between days during the week. We checked a sample of belts coming off the line. They are all within tolerances. Our production mean, median, mode, range, and sigma are very close to a perfect match with the engineers' curve. I'd say this problem is solved."

Peter thanks everyone for their help. He says, "The Customer Service Department is going to be a happy department again."

SUMMARY

The team used the problem-solving process to find the solution to Peter's problem. You have observed the Quality Team as it followed the five-part process. Figure 5.7 shows how each team activity fits the problem-solving process.

SOLVING THE PROBLEM

The Parts	The Activities
Cycle 1: 1. See It: Evaluating the problem	(1) Identifying the problem: Customers complain to Peter that belts are breaking on the new floor polisher. Peter asks the team for help. The team (2) brainstorms for ideas, (3) groups the ideas, and (4) chooses populations to look at.
2. Question It: Gathering good information	The team check their populations: (1) Peter's customers are using the belts correctly. (2) Connie's staff is selling the right belt. (3) Roy gets a week's sample of belts that break too soon by shift and by day.
3. Think about It: Analyzing the data	(1) Data shows that significantly more belts break at the end of the week. (2) It shows no difference among the shifts.
4. Aim at It: Predicting the cause	Ang-Sun thinks the machines might not be working well by Thursday. She suggests that the machines be adjusted in the middle of the week.
5. Score It: Evaluating the change	Terry adjusts the machines on Wednesday. Another sample is done. It shows the same thing: More belts break on Thursday and Friday. The change does not solve the problem. (Back to part 1)

FIGURE 5.7
The teams's problem-solving process

SOLVING THE PROBLEM

The Parts	The Activities
Cycle 2: 1. See It: Identifying the problem	The problem has not changed. George suggests checking the material.
2. Question It: Gathering good information	The team gathers a new sample of 200 belts and measures their thickness.
3. Think about It: Analyzing the data	1. The team measures 40 belts a day for a week. They do a tally of belt thicknesses 2. The team checks the tolerance of the belt specs and compares their sample to the engineers'. They find many belts are not within tolerance.
4. Aim at It: Predicting the Cause	1. The team thinks that Famous Fabric is causing the problem with their thinner fabric and that Richardson is posing a potential problem with their thicker fabric. 2. Both suppliers are told to work within tolerances.
5. Score It: Evaluating the change	Problem solved!

FIGURE 5.7
The teams's problem-solving process (*continued*)

ACTIVITIES

See Appendix A for answers and discussion.

1. Place the letter of the word next to its meaning:
   a. axis
   b. standard deviation
   c. stable process
   d. dispersion
   e. vertical line
   f. horizontal line
   g. sigma
   h. probability

   ___ always doing something in the same way
   ___ a Greek letter that stands for a standard deviation
   ___ the spread of data in a distribution
   ___ one leg of a line graph
   ___ runs across the page
   ___ runs from top to bottom of a page
   ___ the odds that something will happen
   ___ the measure of dispersion

2. Make a problem-solving chart like the one in Figure 5.7. Write the number of each statement below next to the problem-solving part it belongs in. Tell why you think it belongs where you put it. (There may be more than one place you want to use the statement.)
   1. Maria can't get copies of her report.
   2. The copier needs toner.
   3. This is the third time in six months the toner delivery has been late.
   4. Maria's secretary says the copier is not working.
   5. Maria's secretary says, "Maybe one supplier is always late."
   6. Maria asks why the copier is broken.
   7. The secretary checks the storeroom.
   8. The records are checked again, this time by supplier.
   9. The monthly delivery records are checked.
   10. Office Chief Company has had three late deliveries.
   11. Maria buys toner from three suppliers.
   12. All the deliveries are on time for the next six months.
   13. Maria calls Office Chief to ask that they be sure toner is delivered on time.
   14. Records say the toner was ordered two weeks ago.
   15. There is no toner in the storeroom.

3. Use the tally sheet in Figure 5.2.
   a. Make a line graph for Wednesday's data. Use 0's to mark the thicknesses on the graph. Connect the 0's with a solid line.

SOLVING THE PROBLEM

    b. Add the engineers' curve from Figure 5.1 to your line graph.
    c. List three ways Wednesday's data is different from the engineers' data.
    d. How many belts in Wednesday's sample are outside the tolerance limits?

4. Choose a problem in your workplace. Decide how you could use the problem-solving parts to work on it. Write down the steps you could take. Can you solve the problem? Why or why not?

6

Using Line Graphs

■ ■ ■ ■ ■ GOALS
(1) To apply the problem-solving process to a shipping problem.
(2) To use a line graph to show a trend and a cycle.
(3) To record positive and negative data.
(4) To recognize differences caused by changes in scale.
(5) To understand common causes, special causes, and runs.

WORDS AT WORK scattergram, trend, run chart, cycle, variation, run, outlier

Roy is very upset. Three times in the last month, he has had to shut down part of the line. Twice, the line ran out of the oil that lubricates the special cutter for generator belts. Another time there were no new blades to replace worn-out ones. Roy went to Maria's office in the Purchasing Department.

USING LINE GRAPHS

Part 1: See It "Maria, I'm ready to explode. It looks like our just-in-time inventory control isn't working. What's going on here?"

"Hello, Roy, come on in and sit down. I don't blame you for feeling angry. Shutting down a line is very expensive. Let me show you some data I've been gathering over the last month. For some reason, Industrial Machine Supply Company (IMS) has made several late shipments. They're usually very good, so I only spotchecked the orders until I heard about the problem."

"Show me what you have," says Roy. "Frank, Ang-Sun, and Judy have all told me about their problems."

"I've been looking at the number of days it takes to get an order from IMS once it's placed. I've already checked on our staff's 'order-out' time, and it hasn't changed. Our 'order-out' time is well within acceptable range to maintain the quality of our just-in-time program."

"I'm happy to hear that," Roy says. "I know your staff usually does a great job."

Part 2: Question It "Here's the data," says Maria. She shows Roy the chart in Figure 6.1.

"I know we said that orders should get here between 15 and 25 days after the order is placed. That keeps our just-in-time inventory close enough. We don't have to place expensive rush orders or use high-rent storage space. It also gives us a few days before I have to shut down. If I don't get those orders before the 28th day, I shut down somewhere. That costs big bucks!"

SHIPMENT DATA FOR 20 BUSINESS DAYS:
NUMBER OF DAYS FOR ORDERS TO ARRIVE

Business Days	Day 1	Day 2	Day 3	Day 4	Day 5	Day 6	Day 7	Day 8	Day 9	Day 10
Oil	26	30	26	27	26	25	23	24	24	18
Blades	22	26	28	25	25	24	23	21	18	24

Business Days	Day 11	Day 12	Day 13	Day 14	Day 15	Day 16	Day 17	Day 18	Day 19	Day 20
Oil	16	18	14	12	20	14	20	25	23	30
Blades	18	14	15	15	1	15	16	20	23	25

FIGURE 6.1
Maria's chart

"As I look at this month's data, I can see 4 days with deliveries too soon and 6 days with deliveries overdue enough to cause a problem," Maria says. "I guess you've been lucky more than once."

"Luck isn't good enough, and we both know it!" Roy says. "I'm really worried about this. Let's do a **scattergram** to see if there's a pattern or **trend**. Since the same supplier delivers both oil and blades, we can combine the data on one chart."

WORDS AT WORK

scattergram (**skat** er gram)   The picture shown by plotting raw data on a graph. It may show a left-right or up-down pattern, or no pattern.
trend (tr **end**)   A pattern of events that has meaning. It probably has an assignable (explainable) cause.

SCATTERGRAMS

"OK, first we need a graph with a vertical and horizontal axis," agrees Maria. "Let's put the 20 working days of last month on the horizontal axis, and the 'days to receive' on the vertical axis. We placed orders each day for oil and blades with IMS. Let's make the horizontal axis interval equal to one day. That way we can see most clearly what's happening to the day's orders."

"OK, what was the longest time to get an order?"
"Thirty days."
"Do we need to list all 30 on the vertical axis?"

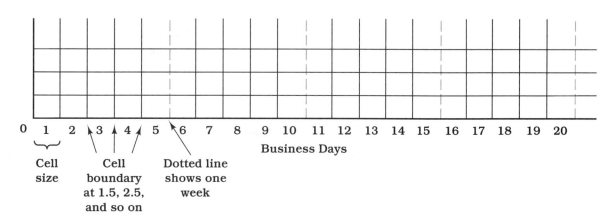

FIGURE 6.2
Shipment graph, horizontal axis

USING LINE GRAPHS

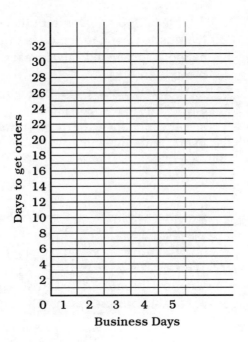

FIGURE 6.3
Shipment graph, vertical axis

"The shortest time was 1 day. We can label the intervals on the vertical axis by twos on every other line."

Help Roy and Maria by making a checkmark opposite the number of days to receive each order. Use the data in Figure 6.1 and the blank graph in Figure 6.4. The first week has been done for you by Roy.

The completed graph is in Figure 6.5. Compare your graph to it. Do they look the same?

 Part 3: Think about It Where do you see open spaces on the graph? You are right. There are no checks in the lower left part of the graph (below 20). There are no checks in the upper right of the graph (above 22) between days 10 and 17. Do the checkmarks seem to follow a path from day 1 to day 20? What could you do to make the pattern clearer?

Moving from Scatter to Line

Remember in Chapter 4 we talked about measures of central tendency. Let's see what happens when we plot the averages (means) for each day on the graph in Figure 6.5.

Practice: Make a copy of Figure 6.5 and use the data in Figure 6.1 to plot the averages for each day. Place a small X on your graph where each day's mean belongs. For example, the

■ 66 ■

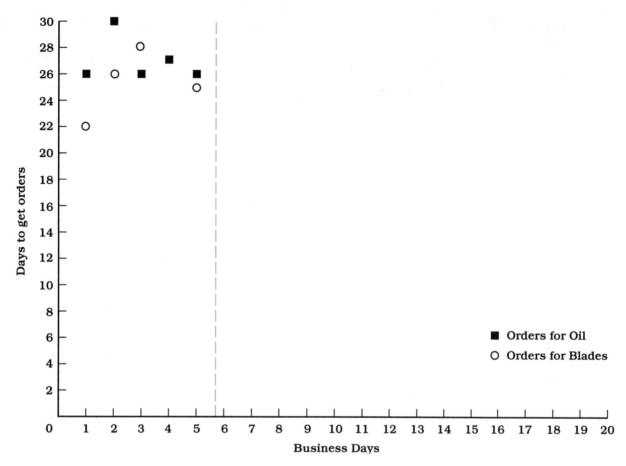

FIGURE 6.4
Shipment graph, week 1 filled in

mean for day 1 is 24. Put an X at 24 on the day 1 line. Complete your graph. Check your answer in Appendix A.

Draw a line with a ruler from X to X. What can you tell now about the path from day 1 to day 20? During what weeks were there days whose orders took 25 or more days on average? That's right, days 2, 3, 4, 5, and 20 had daily averages of 25 or greater. Days 13, 14, 15, and 16 had averages below 15.

Line graphs are used to show trends, how things look over time. Roy and Maria now see that orders placed from day 1 to day 10 and from day 18 to day 20 take longer to receive than orders placed from day 11 to day 17. What day had the highest average? The lowest? Check your answers in Appendix A.

The day with the highest average is in a week of high averages. The day with the lowest average is in a week of low averages. That is another way of making sure that Maria and Roy are looking at a trend.

USING LINE GRAPHS

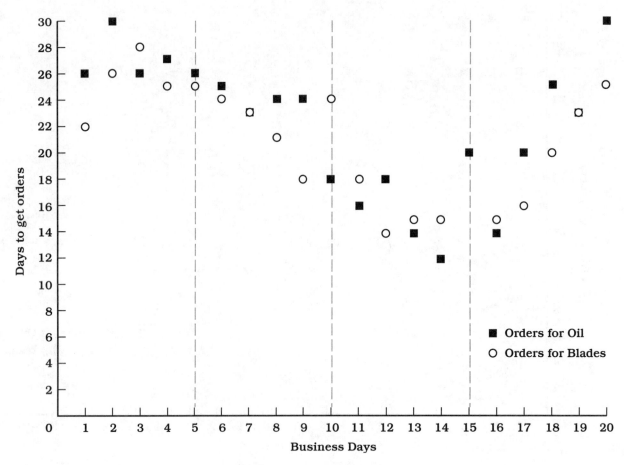

FIGURE 6.5
Shipment graph, weeks 1 to 4 filled in

Changing the Scale on Line Graphs

Another way to make a trend stand out is to change the scale on the axes. If the vertical scale is set at 5, what would the graph look like?

Figure 6.6 shows what happens when each day's data in Figure 6.1 is averaged and plotted on the graph. The vertical scale is also changed to 5 days. Is the graph's pattern more clear or less clear with this scale? There is still a clear empty space in the left lower part of the graph.

Compare Figure 6.6 with Figure 6.5. Do both figures show the same trend? In which weeks did orders take the longest average time? In which weeks did they take the shortest time?

Suppose Maria had just used the average for each week. That would be a scale of 4 points (weeks) on the horizontal

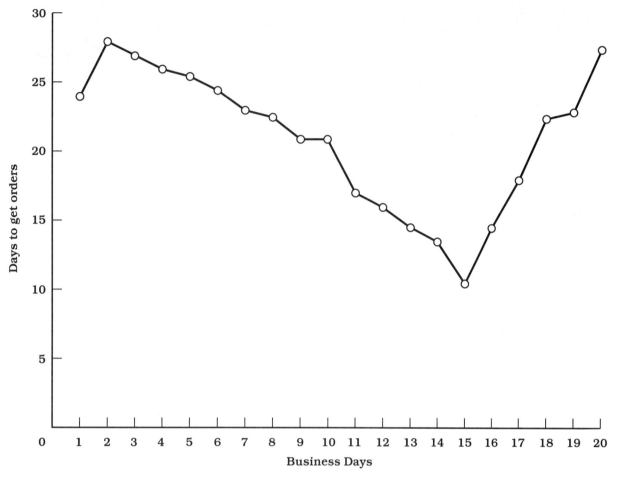

FIGURE 6.6
Shipment data with vertical scale of 5

axis instead of 20 points (days). If she set the interval at 5 for the vertical axis and at five days (one week) for the horizontal axis, her data would look like Figure 6.7.

Does this data also show the longest and shortest receiving weeks? Are they the same weeks shown in Figure 6.6?

When Maria just wants to know the longest and shortest order times of the month, it's all right to estimate (round off). Will Figure 6.7 give her enough information to solve her problem? Let's see. At this point Roy and Maria will backtrack to Part 2: Question It, to gather more information before coming to any conclusion.

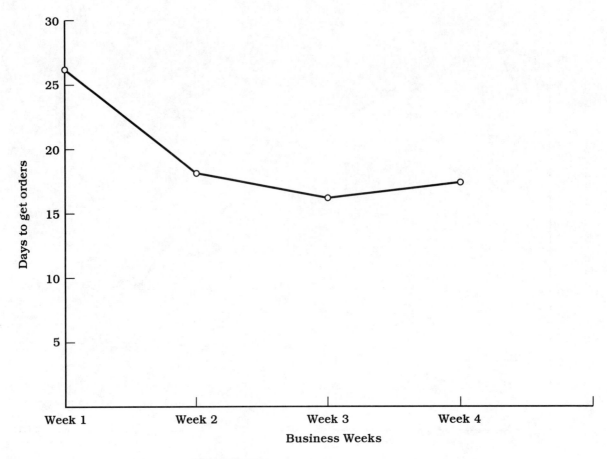

FIGURE 6.7
Shipment data with changed scales

CYCLES Roy asks, "Is this an ordinary month? I mean, does it always take longer to get things at the first and last parts of the month?"

"I wanted to know that, too," Maria answers. "I asked my staff to go back three months and plot the daily averages. The trend holds true. Every month things ordered in the third week get here quickest. The slowest deliveries are right before and after the first of the month."

Finding a Cycle Checking data for three more months helps Maria see patterns. Her four-month data, plotted on a **run chart,** show a pattern called a **cycle.** Each month's graph looks about like the one before and after it.

USING LINE GRAPHS

WORDS AT WORK

> **run chart** A graph showing how a process works over time. It may show upper and lower control limits. It will show when a process is out of control. It can help to find the cause.
> **cycle** (sy kl) A pattern that repeats itself over and over. For example, the seasons of the year and the temperatures that go with them.

To be sure it is really a cycle, data has to be collected for some time. For example, say you keep daily high temperatures for a town in Illinois. You do it for two years and see a cycle. Highs will get higher in the spring, be highest in the summer, go down in the fall, and be lowest in the winter, then repeat the cycle. If you took only temperatures for March, April, and May, what might you conclude from the upward trend? It is important to graph the process long enough to be sure about what is happening.

"But I have not shut the line down before. If the cycle happens every month, why are we having to shut down now?" Roy wonders.

"That's because the delivery times have been within the control limits we set. The run chart shows a clear cycle, but until now, it has not been significant. We always got stuff on time or early."

The Meaning of Cycles

Finding a cycle or other pattern in a graph doesn't necessarily mean something is wrong. Slower delivery times at the ends of the month and faster ones in the middle may just be part of the company's system. There may be many causes. Sometimes no cause can be found.

Variations (changes) in delivery time during the month are a normal part of the process. The difference between day 1 and day 17 does not matter as long as the "longs" and "shorts" stay within the limits the production line needs. Variations that are caused by the system itself have what are called "common" causes. Variations or changes that are not caused by the system itself have what are called "special" or "assignable" causes. Maria and Roy are looking for special causes of late orders.

WORDS AT WORK

> **variation** A change in a process from what is expected. Every process has some things that change while it is going on.

USING LINE GRAPHS

THE IMPORTANCE OF SCALE AND INTERVAL

If Maria had used Figure 6.7, it would have shown the long-short time cycle. However, it would not have shown the days whose orders took more than 25 days. It does not show any orders taking less than 15 days (coming too soon). Remember, we rounded the numbers to the nearest 5. Why is it important to know which orders took more than 25 days? Less than 15 days?

RUN CHARTS

Roy says, "Let's look at your graph as a run chart. That way we can see exactly when the orders were out of control."

"Good idea!" Maria agrees. "Just draw a dotted line across from 15, the shortest delivery time we want. Draw another dotted line across from 25, the longest time we can wait for an order. Do the same thing to the other three month's graphs. Look at this graph now!" (See Figure 6.8.)

"How many days had at least one order too late or too early?" Roy asks. "Days 1, 2, 3, 4, 5, and 20 all had at least one late shipment. Look at that first week!" Of the first 6

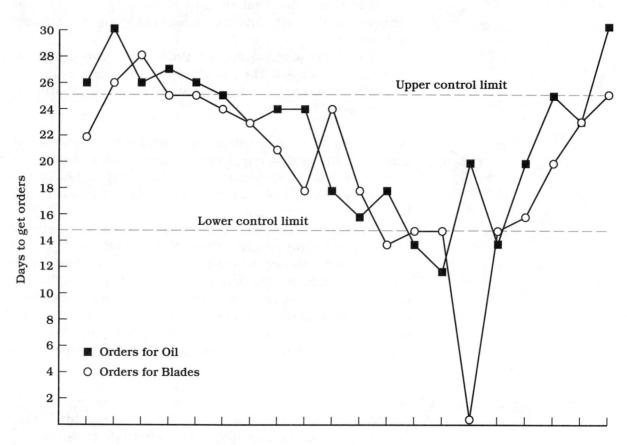

FIGURE 6.8
Shipment graph as a run chart, showing control limits

USING LINE GRAPHS

days, 4 averaged above 25 days. That's a **run**. Day 20 was above 25, too. Only one day, 15, was far below 15 on average, although days 13, 14, and 16 were a day or so too soon. If you look at day 15, you will see that the blades were delivered only one day after they were ordered. That is very unusual. Sometimes an event happens only once or twice in a large number of events. It is called an **outlier**. Outliers (because they "lie" outside of the normal range) are sometimes not even counted in finding an average.

WORDS AT WORK

> **run** graphed data for a series of events that shows a pattern over time.
>
> **outlier** (**out** lie er) an event or piece of data that stands alone far outside the normal range. It is often not counted in the average because it moves the average too far away from the median.

"The blades order was late or early only twice by a day or two. Those one-time events are incidents or occurrences. The averages were in control," says Maria. "I'm not so upset about the early shipments. Early is not great, but it doesn't break us. We don't have to shut down."

GRAPHING DEVIATION FROM THE RANGE MIDPOINT

Another way to show a run chart can be seen in Figure 6.9. It looks very similar to Figure 6.5 except for the vertical axis. The same data is plotted using the midpoint of the range of acceptable delivery days as a zero baseline. Remember, Roy said he wanted supplies delivered between 15 and 25 days from the order day. The midpoint of that range is 20 (25 + 15 = 40 ÷ 2 = 20). Figure 6.9 shows the 20th day as zero. The 25th day is labeled +5. The 15th day is labeled –5. Recording data this way shows an incident or a run very clearly. Everything above +5 or below –5 is out of control.

If Maria kept shipping data every day for a year, she could make a normal distribution curve as the team did in Chapter 5. However, this is a short-term problem. Using a run chart gives Roy and Maria all the data they need.

 Part 4: Aim at It "I'm glad you used a small interval scale," Roy says. "If you had used too big an interval on the graphs, we might have missed that run. What do we do now?"

"I'll call IMS right away. Now I have some good data for the four months. I can tell them we know they are usually

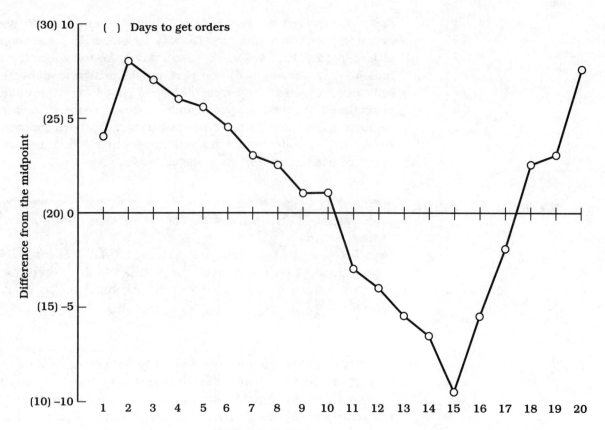

FIGURE 6.9
Shipment graph with data plotted around range midpoint

right on target with deliveries. There must be a special cause for the delays this month. I'll call you as soon as I have any information," Maria promises.

"Great! I'm going to check my inventory on the computer right now. If it looks like I'll be too close for comfort with any parts or supplies from IMS, I'll call you immediately," says Roy as he leaves.

Part 5: Score It Later, Roy and Maria talk on the phone. "Roy, hi," says Maria. "I talked to Sandy at IMS. He says it has been a real case of Murphy's Law lately: Whatever can go wrong, will. They had a bad truck accident. The whole truckload was destroyed by fire. Then, two shipping handlers got the flu. Last, and worst, remember that bad thunderstorm we had last month? It knocked out their computers for eight hours. Sandy apologized. He will rush-order at no extra charge whatever we need for next week. That way we will have enough until he gets the mess cleared up."

USING LINE GRAPHS

"All right!" Roy says. "Maria, you're great! Here's the list of what I'll need . . ."

SUMMARY

Maria and Roy used a scattergram and a line graph to look for a trend. They needed to find out why orders were coming in late. They collected data and plotted it on a graph. They found the average for each day, and plotted those. The line graph of averages showed the trend more clearly. Deliveries were slower at the beginning and end of the month. Roy and Maria wondered if that was always true. They checked data for three previous months. The trend really was a cycle: it happened over and over again. However, this month the cycle was out of control. Other months, it was in control. Maria and Roy knew the cycle was not the cause of the problem. The cycle was a common cause. They were looking for a special cause, something that could be fixed.

They changed scales to see if that made things clearer. They also made a run chart to see exactly when deliveries were out of control. Maria and Roy know there is more than one way to show the data. One way is a control chart showing the average delivery time for each day. On this chart, a line is drawn to show the shortest and the longest acceptable delivery days. Deliveries within the lines are in control. Deliveries outside the lines show problems. Another way is to graph the data around the range midpoint which is set at zero. Acceptable deliveries are between +5 and –5. Problem deliveries will show up as more than +5 or less than –5.

Maria called IMS and learned about the causes: an accident, an illness, and a computer failure. These are special causes. They are not basic to the delivery system. Sandy gave Best Belt a rush delivery at no extra cost. Roy is happy because he will have enough supplies to run the line. Maria has solved a serious supply problem. Sandy has made an unhappy customer happy again.

Knowing the difference between a common cause and a special cause allows Roy and Maria to solve their problem.

ACTIVITIES

Answers and assistance may be found in Appendix A.

1. On page 64, Roy says, "I know we said that orders should get here between 15 and 25 days after the order is placed." Using this information, draw the upper control limit line on your graph (Figure 6.4). Draw the lower control limit.

USING LINE GRAPHS

   a. List the delivery-day averages that are too short. Find the mean.
   b. List the delivery-day averages that are too long. Find the mean.
   c. List two reasons why delivery-day times might be too short. Why could that be a problem?

2. Use the data below to create a scattergram by following steps a to f below.

   Number of Customer Service Calls

Monday	Tuesday	Wednesday	Thursday	Friday
12	13	12	10	11
13	13	12	11	11
15	10	11	12	12
17	10	10	11	9
14	15	12	10	8
14	14	11	8	8
9	15	8	10	12
12	12	11	12	10
13	14	13	12	10
13	11	15	12	11

   a. Label the vertical and horizontal axes. (Hint: It's easier if you make the number of calls the vertical axis.)
   b. Use a checkmark to show where each number fits on the graph.
   c. Compute each day's mean and plot it on the graph with an "o".
   d. Draw a line through the means.
   e. What is the trend?
   f. Can you tell if there is a cycle? Why or why not?

3. Use an activity which happens frequently in your workplace (phone calls, employees absent, orders received or shipped, etc.).
   a. Record at least 20 values for your activity (time of call, number absent, number of orders, etc.) for each day for three days.
   b. Make a scattergram.
   c. Find the mean for each day/hour. Plot it on the graph.
   d. Can you see a trend? If so, what is it? Is there a cycle? If so, what is it? If not, why not?

7

Using Bar Graphs

■ ■ ■ ■ ■ *GOALS*
**(1) To set up and read a bar graph showing frequency data.
(2) To review a bar graph showing a normal distribution, a skewed distribution, and a bimodal distribution.**

WORDS AT WORK skewed, bimodal

Connie Borrelli is working on the Marketing Department's advertising budget for next year. She wants to be sure she spends the money only on ads that bring business to Best Belt.

Everyone who speaks to customers asks, "How did you find out about Best Belt Company?" Connie's secretary keeps a tally of the answers. Figure 7.1 shows her data for last year in Column 4.

Connie wants to show the other members of the team her results and her plan for next year. Watch her use the problem-solving process to find out what Figure 7.1 tells her.

■ 77 ■

USING BAR GRAPHS

Col. No.	Col. 1	Col. 2	Col. 3	Col. 4	Col. 5	Col. 6	Col. 7
	Number of Ads	Cost per Ad	Total Cost	Number of Responses	Cost per Response	Income per Response	Total Earned by Advertising
Newspaper	10	$ 200	$ 2,000	40	$ 50	$2,000	$ 80,000
Trade journal	20	400	8,000	100	80	2,500	250,000
Press release	6	0	0	10	0	3,000	30,000
Radio	104	165	17,160	20	858	1,000	20,000
Commercial TV	4	3,000	12,000	100	120	3,500	350,000
Public TV	50	250	12,500	100	125	3,000	300,000
Total	194		$51,660	370	$1,233		$1,030,000
Average		$669.17			205.50		

FIGURE 7.1
Connie's new customer data

Problem Solving Part 1: See It Connie's budget for last year was $51,660. Her boss has asked her to cut her budget by 20 percent next year. Problem: What kind of advertising should she drop?

Problem Solving Part 2: Question It Connie has the data she needs. If she makes a bar graph, she can see more clearly. First, she decides to see which kind of ads brought in the most money for each response. She sets up the graph with the vertical axis showing the income per response. She decides to use a cell size (interval) of 500. Figure 7.2 shows what the vertical axis will look like. The horizontal axis will be as shown in Figure 7.3.

Practice: Make a graph like this on a sheet of paper. Use the data from column 6 in Figure 7.1 to plot the income per response for each kind of advertisement. Make a bar from each dot to the horizontal axis. The bars should be equal width, roughly as wide as the word describing the kind of advertisement. Your graph should look like that in Figure 7.4. Now you can see very quickly which ad was best—commercial TV.

Problem Solving Part 3: Think about It Look at column 2 of Figure 7.1. Which ad is most expensive? Right, commercial TV. Which is least expensive? Press releases are free. Does

■ 78 ■

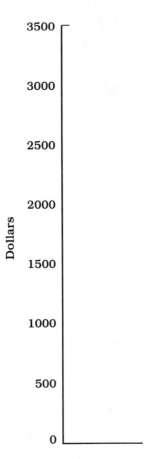

FIGURE 7.2
Vertical axis of the bar graph

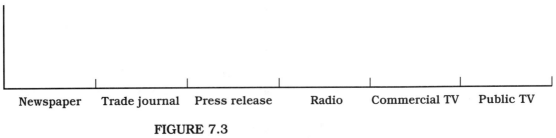

FIGURE 7.3
Horizontal axis of the bar graph

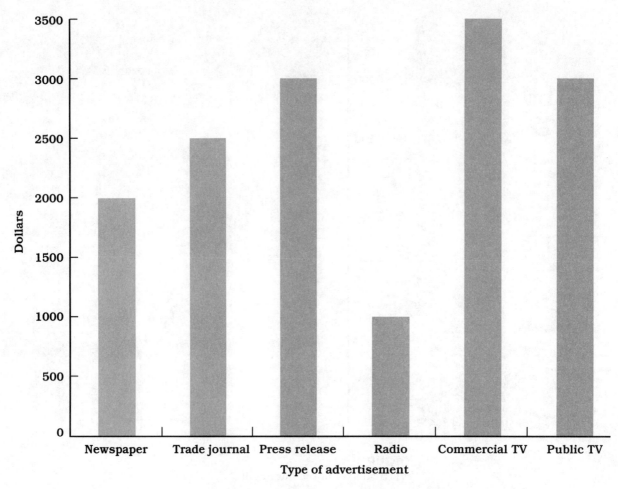

FIGURE 7.4
Connie's bar chart of income per response

that mean press releases are the best advertisements to use? Why or why not?

Connie knows that people who respond to the ads really buy from Best Belt. Her data shows her the total amount earned from each type of advertising. She graphed the total income for each type of ad. Which column in Figure 7.1 did she use? Right, column 7.

Practice Graph column 7 as you did column 6. (Hint: If you set the vertical scale at intervals of 25,000 it will look like Figure 7.5.)

Using a graph to compare two sets of data about each category (kind) of ad can be helpful. Connie takes the "cost per

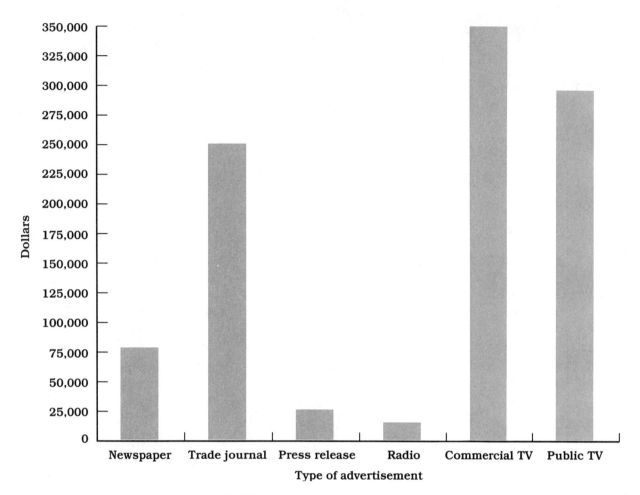

FIGURE 7.5
Total income from advertising

response" and the "income per response" columns and creates the graph in Figure 7.6 on page 82. She can now see quickly that one category stands out. The bars for radio ads show that each response cost almost as much as it brought in. What do you think the team will say about using radio ads next year?

USING BAR GRAPHS TO SHOW DISTRIBUTIONS

In Chapter 2 you learned about distributions. A bar graph can show differences in distributions very well. Figure 4.4 on page 43 is a bar graph of the engineers' sample of belts in Chapter 4. Turn Figure 4.4 on its side. The frequency distribution looks like a bar graph. The bar graph shows a normal distribution just as the bell curve in Figure 5.1 on page 48 does. If

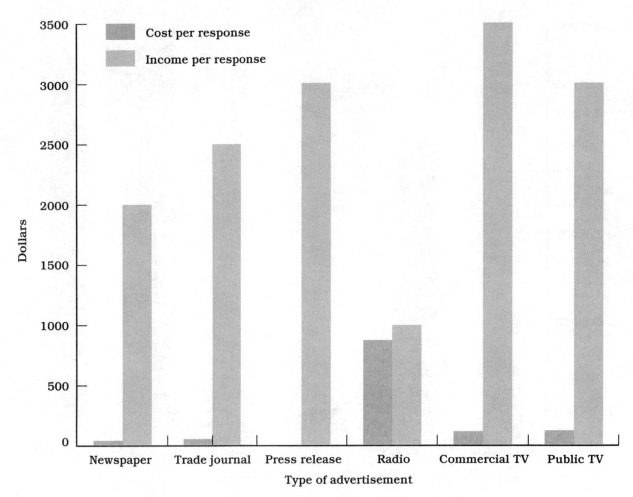

FIGURE 7.6
Comparison of cost per response and income per response

you draw a line above the bars, you will have reproduced the bell curve in Figure 5.1.

Bar graphs, like distribution curves, can also show **skewed** distributions. Look at Figure 7.7. It is a bar graph of the cost of each response by category. The bar labeled "average" shows the average cost. The graph shows that radio is far above the average cost.

WORDS AT WORK

skewed (s qued)   Off center, out of balance, having more at one end than at the other.

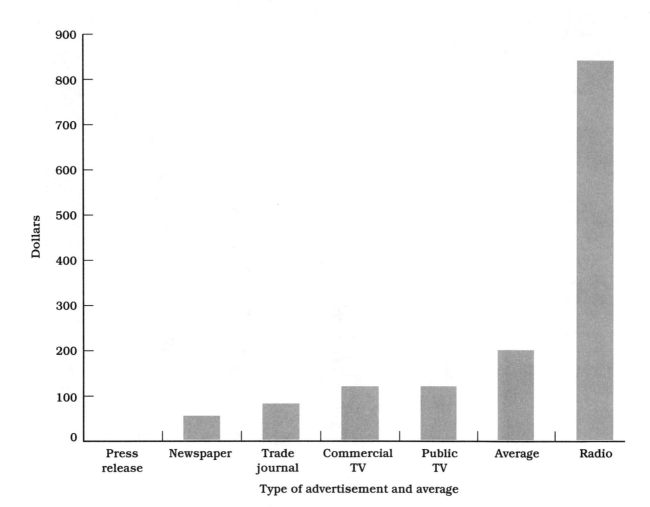

FIGURE 7.7
Cost per response compared to average

Practice: Figure the average cost without including radio. What did you find? What does that number tell you about radio advertising? Figure 7.8 gives you the picture in bar graph form.

Without radio, the distribution is less skewed. The average in Figure 7.8 is not a good measure of the midpoint of the cost. Distributions can be skewed when some data is much greater or much less than the rest. In a quality or process control situation, it is important to know why a distribution is skewed.

There is another way Connie can show the team her data about new customers. She could use the average of the total cost of ad types, labeled zero, as the horizontal axis. The

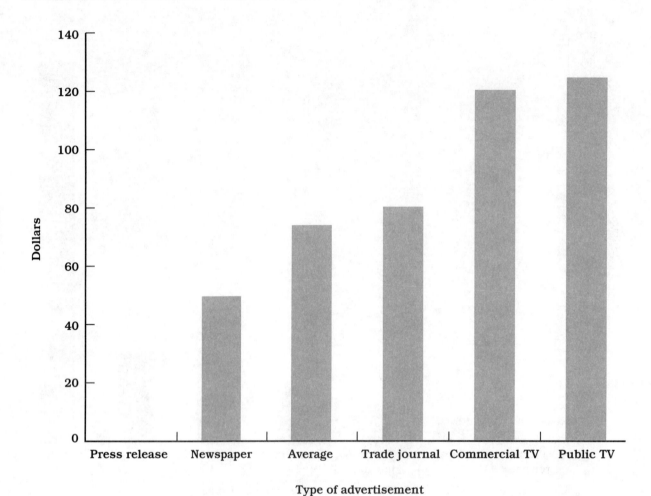

FIGURE 7.8
Cost per response compared to average excluding radio

vertical axis will go above the zero (positive) and below the zero (negative). Then, she will subtract each cost amount from the average, $8,610. She will get some negative numbers. That's because some of the types of ads show smaller amounts than the average. When Connie graphs this data, her picture will look like Figure 7.9. Does that graph give you a clear idea of the problem with radio advertising? Its cost is far above average.

BIMODAL DISTRIBUTIONS Another distribution you might see is called a **bimodal** distribution. If data is grouped around two points, your graph might look like the one in Figure 7.10.

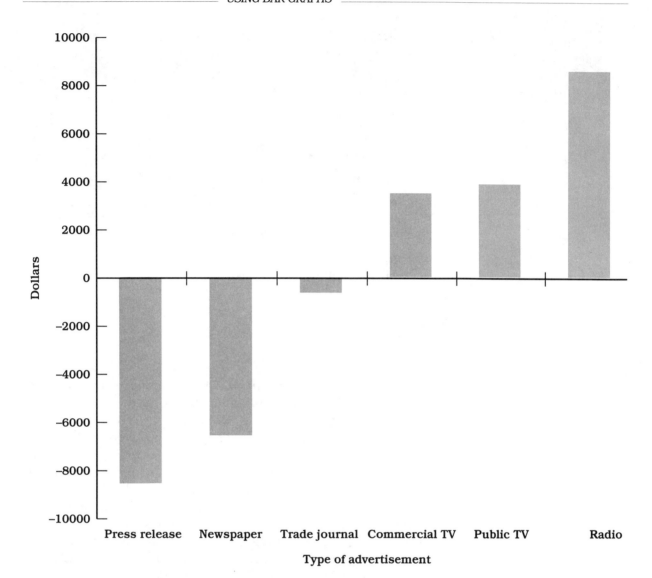

FIGURE 7.9
Total costs of each type of advertising in relation to the average

WORDS AT WORK

bimodal (by mow dal)   Having two modes. On a graph, the distribution has two high points. There are fewer items at the mean (average) than at the high points.

This is what you would see if there had been 34 belts 8.25 millimeters thick and 34 belts 14.25 millimeters thick. If the median had been 3, the engineers' distribution would be very different. Compare this figure with Figure 4.4 on page 43.

USING BAR GRAPHS

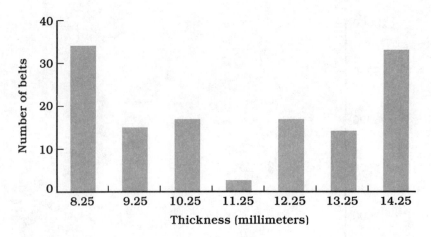

FIGURE 7.10
Bimodal distribution of belt thicknesses

As you learned in Chapter 4, the mode is the value of data that happens most often. In the following distribution, there are two modes: 2, 2, 2, 3, 3, 3, 3, 4, 5, 5, 5, 6, 6, 6, 6. Both 3 and 6 have four items. They are both modes. Turn this data set into a bar graph. It will look similar to Figure 7.10. Bimodal distributions most often happen when you compare data sets that are different. For example, we talked about measuring height at a Super Bowl game. If we put everyone in the same sample, the distribution should be normal (a bell curve). Divide the group into men and women, measure them separately, and graph the results, and most likely you would get a bimodal distribution. Why? Men are taller on average than women. The mode for men will be different than that for women. Each population will have its own mode.

What do you think Connie will do in parts three, four, and five of the problem-solving process? In the next chapter she will create some circle graphs to show the team.

SUMMARY Connie made a table of data about this year's advertising budget. She used the table to make bar graphs. That helped her see that radio is not a good form of advertising for Best Belt. Her bar graphs showed that radio costs more than the average per response. In fact, responders to radio ads almost did not buy enough belts to cover the cost of the ads. When she sets up next year's budget, do you think she will put radio in it?

ACTIVITIES   Answers and assistance may be found in Appendix A.

1. Use the information in column 1, column 2, and column 3 of Figure 7.1 to set up three bar graphs.

2. Use the average of column 2 to make a zero-based graph for the cost per ad. Be sure to keep your vertical axis intervals equal. It will be easier if you use graph paper. Remember, you have to subtract the cost of each ad from the mean to find the number to plot on the graph.

3. Write Connie's problem-solving process in a copy of the chart in Figure 5.7 on page 58. Fill in what you think she will do in parts three, four, and five.

Using Circle or Pie Graphs

■ ■ ■ ■ ■ *GOAL*
To use circle graphs to show the relationship of parts to the whole.

WORDS AT WORK   protractor, compass, diameter, radius, circumference

In Chapter 7, Connie gathered data on advertising and its results for this year. This data is duplicated in Figure 8.1. She learned that her budget for next year will be 20 percent less than she spent this year ($51,660). (A quick review of percents is in Appendix B.) To figure out how much money she will have next year, she follows these steps:

1. Turn 20 percent into a decimal (.20).
2. Multiply the total budget for this year, 51,660, by the rate, .20 (51,660 × .20 = 10,332).
3. Subtract the budget cut (10,332) from this year's budget (51,660 − 10,332 = 41,328).

Connie will have $41,328 to spend on advertising next year.

USING CIRCLE OR PIE GRAPHS

Col. No.	Col. 1	Col. 2	Col. 3	Col. 4	Col. 5	Col. 6	Col. 7
	Number of Ads	Cost per Ad	Total Cost	Number of Responses	Cost per Response	Income per Response	Total Earned by Advertising
Newspaper	10	$ 200	$ 2,000	40	$ 50	$2,000	$ 80,000
Trade journal	20	400	8,000	100	80	2,500	250,000
Press release	6	0	0	10	0	3,000	30,000
Radio	104	165	17,160	20	858	1,000	
Commercial TV	4	3,000	12,000	100	120	3,500	20,000
Public TV	50	250	12,500	100	125	3,000	350,000
Total	194		$51,660	370	$1,233		300,000
Average		$669.17			205.50		$1,030,000

FIGURE 8.1
Connie's new customer data

DETERMINING PERCENTAGE OF TOTAL

Connie wants to spread the cuts among the types of advertising. She does not want to give up ads that work. She decides to find out what percent each kind of ad is of this year's budget. For example, what part of $51,660 is the cost of newspaper ads ($2,000)? To find out, divide 2,000 by 51,660. Right, the answer, rounded, is .039 or 3.9%.

Practice: Complete the chart in Figure 8.2. It's faster if you have a calculator. Round to the third decimal place, then move the decimal point two places to the right to change it to a percent.

Did you notice that 0 divided by 51,660 is still 0? Since press releases don't cost anything, they have no share of the

	Cost	÷	Total	=	Percent
Newspaper	2,000		51,660		3.9%
Trade Journal	8,000		51,660		_____
Press release	0		51,660		_____
Radio	17,160		51,660		_____
Commercial TV	12,000		51,660		_____
Public TV	12,500		51,660		_____

FIGURE 8.2
Total cost of ads by type, as percentage of total

cost budget. Does that mean they don't have a share of the income? Do your answers add up to within one percentage point of 100 percent? If not, something is wrong. (Due to the loss or addition of fractional amounts in rounding, percentages of the total may add up to slightly more or less than 100 percent.) Check your answers with Figure 8.4.

CIRCLE GRAPHS

Circle graphs (also called pie charts) show data that add up to 100 percent. They show what share of the whole each piece is.

When the team sees Connie's circle graphs, they will be able to compare data easily. Connie will make a circle graph of the cost and one of the income from all the types of ads. Putting them side by side will make the comparison clear.

Today, computers often make graphs for us. You can make a circle graph with a few simple calculations. The tools you need are a **compass** and a **protractor**.

FIGURE 8.3
A compass (© John Greer/UNICORN STOCK PHOTOS)

USING CIRCLE OR PIE GRAPHS

WORDS AT WORK

> **compass** (**com** pass) A tool for drawing circles, with a sharp point on one arm and a pencil on the other. The two arms are joined at the top and crossed by a curved "ruler." The ruler determines the radius and thus the size of the circle.
>
> **protractor** (**pro** track tore) A tool which measures angles and circles in degrees. It looks like a half circle.

Making a Circle Graph

Follow the steps Connie uses to make a circle graph by hand.

Problem Solving Step 1: Make a circle with a compass. Pull the compass apart until the bar indicates a **radius** of 3. That will give you a circle with a **diameter** of 6 inches. Place the point in the middle of the paper. Hold it firmly and swing the pencil around it until the circle is formed. Circles always have 360 degrees. Degrees are units of measure like inches and feet. They are used to measure curved surfaces and angles. Connie can use them as if they were inches.

WORDS AT WORK

> **radius** (**ray** dee us) The distance or the line from the center of a circle to anywhere on the edge. The radius is one-half the diameter. In a circle graph, the space between the radius lines shows the shares.
>
> **diameter** (die **am** eter) The line that cuts a circle in two equal pieces. Circles are usually named by their diameter. A six-inch circle has a diameter of six inches.

Problem Solving Step 2: Convert the percentages into degree shares. Connie knows that 100 percent is all there is. If a circle equals 360 degrees, then 360 degrees is 100 percent of a circle. The circle (or pie) graph will be divided into "slices" based on the percentages converted into degrees. The cost of a type of ad as a share of total cost, its percentage of 100 percent, and its share of degrees out of 360 degrees all are equivalent.

Connie needs to change the percentages you found in Figure 8.2 into degrees so she can divide up the circle. Because the relationship is the same, she can use the same percentages.

She sets up a table. She knows the percentages (shares) and the total number of degrees (360). To find the number of degrees for each type of ad, she multiplies 360 by each of the percentages. Complete the chart in Figure 8.4. Use a calcula-

USING CIRCLE OR PIE GRAPHS

	Cost of Ad	Percent	×	Total	=	Degrees
Newpaper	2000	3.9%		360		—
Trade journal	8000	15.5%		360		56
Press release	0	0%		360		0
Radio	17,160	33.2%		360		119*
Commercial TV	12,000	23.2%		360		84
Public TV	12,500	24.2%		360		—
Total	51,660	100%				360

*The full answer, 119.52, has been rounded to 119 instead of 120 to make the numbers total 360 degrees.

FIGURE 8.4
Total cost of ads by type, as degrees of a whole circle

tor if you can. Round off to whole degrees. See the correct answers in Figure 8.6.

You remember how you calculated Figure 8.2. You divided the cost of each ad type by the total cost to get the percent. This time, you know the percent and the total degrees, so you multiply the total by the percent to get the degrees per ad type. Now Connie can create the graph.

Problem Solving Step 3: Mark the degree shares on the circle. Connie uses the protractor to mark the circle with the number of degrees for each ad type. It helps to draw a dotted line across the center of the circle. Put the center of the protractor on the spot where the compass made its hole. Line the base of the protractor up with the dotted line. Work from the left and put a mark at 14 degrees on the protractor arch. Draw a line from the center dot to the **circumference** of the circle. It should look as shown in Figure 8.5.

WORDS AT WORK

circumference (sir **come** fur enss) The round edge of the circle. It can be measured in degrees and always has 360 degrees, no matter what size it is.

Connie can enter the next point by placing the protractor on the 14-degree line as if it were 0. Then she can measure 56 degrees from the 14-degree line and make another point with a line. Make the rest of the measurements and draw the lines. The last line should fall on the dotted center line, back where

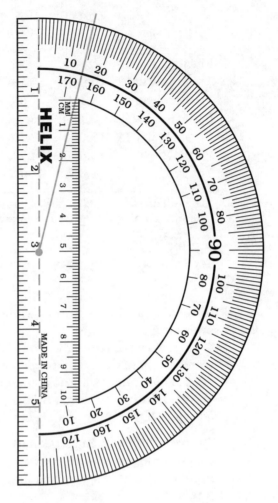

FIGURE 8.5
Using a protractor to mark a center line and 14 degrees from it.

you started. Your graph should now look like that in Figure 8.6. After it is labeled and shaded it can look like that in Figure 8.7.

Each pie- or pizza-slice shaped piece shows how the cost of one type of ad relates to the whole cost. Does the circle graph make it easy to see the most and least expensive ads?

Comparing Circle Graphs   Connie uses her computer to make another circle graph. This time she graphs column 7 of Figure 8.1, the total earned by advertising. The computer will use the data to create the chart shown in Figure 8.8. It will then create the graph shown in Figure 8.9. Does this graph make it clear which kind of advertising earns the most? Did you notice that press releases have a share in this graph? Why?

USING CIRCLE OR PIE GRAPHS

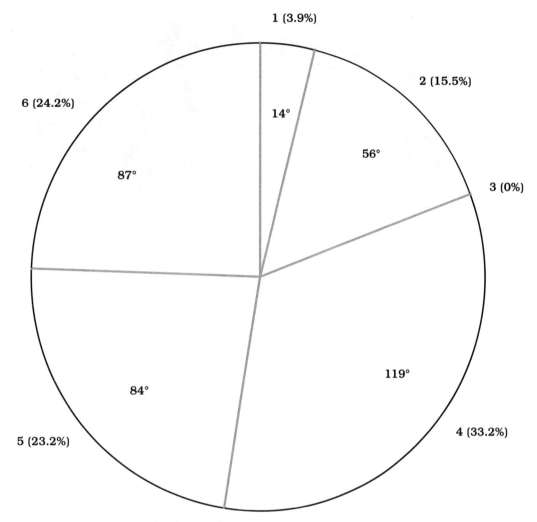

FIGURE 8.6
Marking the sections by measuring degrees

The next day, Connie meets with the team to continue part three of the problem-solving process: Think about it. She has the bar graphs from Chapter 7 and these circle graphs.

"As you know, the advertising budget has to be cut 20 percent next year," Connie says. "That means I have $10,332 less to spend. Next year's budget will be $41,328. This bar graph compares advertising cost and income per response for each type of ad we used." She shows them the bar graph from Figure 7.6 on page 82.

Peter says, "That chart certainly shows that press releases are the way to advertise! Look at how much revenue there is compared to expense."

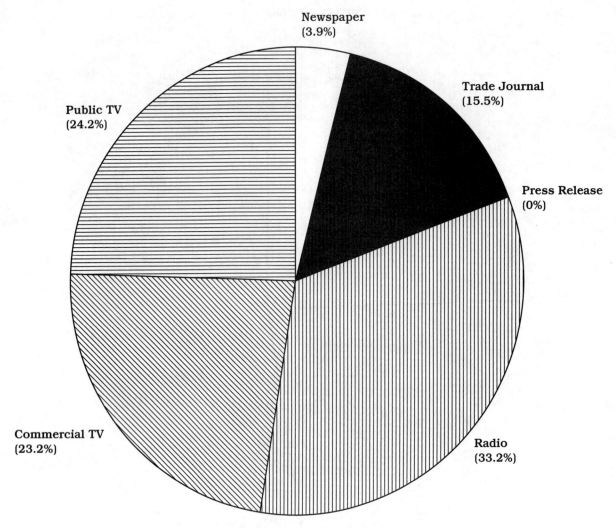

FIGURE 8.7
Connie's circle graph of costs

	Total Earned by Advertising	Percentage of Total	Number of Degrees
Newspaper	$ 80,000	7.8%	28
Trade journal	250,000	24.3%	88*
Press release	30,000	2.9%	10
Radio	20,000	1.9%	7
Commercial TV	350,000	34.0%	122
Public TV	300,000	29.1%	105
Total	1,030,000	100%	360

*The full answer, 87.48, has been rounded to 88 instead of 87 to make the numbers total 360 degrees.

FIGURE 8.8
Data for income graph

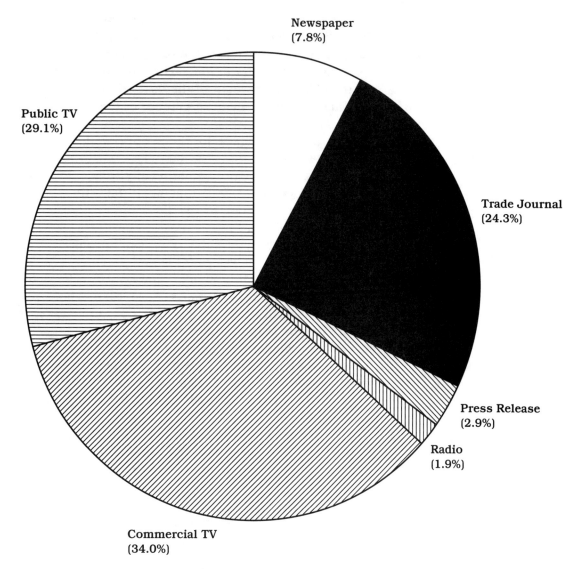

FIGURE 8.9
Connie's circle graph of income

"Yes, but look at radio," says Roy. "Those ads cost almost as much as they brought in."

Maria agrees, "If I had to cut, I'd sure cut radio first."

Connie shows the graph in Figure 7.8. "It's clear to me that the cost of radio ads is far above average."

"What percent of this year's budget was radio advertising?" asks Terry. "That might be a good way to look at how to cut. Maybe all you need to cut is radio."

"I don't think that will work," says Connie. "Radio ads bring in $20,000. I don't think we want to lose all that income."

Roy says, "Let's see what the graphs tell us. First, we already know that radio costs a lot. Radio is 33 percent of the cost and only 2 percent of the income. That's the worst."

"Yes," says Maria, "television is clearly the best. Commercial TV is 23 percent of the cost and creates 34 percent of the income. Public TV is 24 percent of the cost and 29 percent of the income."

Peter adds, "Trade journals work well, too. What about press releases?"

"Press releases don't cost us anything, and they do bring in almost 3 percent of our income. In fact, they bring in more than radio at no cost to us. I'd want to continue them for sure," says Connie.

"Connie, can you show us a way to look at next year, if everything is cut 20 percent?"

"Yes, look at this chart [Figure 8.10]. These bars show this year's costs less 20 percent for each type. I used a bar chart. If everything is 20 percent less the shares stay the same. A circle graph won't look any different."

"The problem is my cost per ad will not go down just because my budget does. I need to see how to use my money wisely," Connie says. "I won't be able to buy as many ads. That means we may not get as much income from advertising."

Figure 8.11 shows the number of ads Connie can buy next year at 20 percent less than this year. Since Connie can't buy a part of an ad, you can round the numbers to the nearest whole ad.

Problem Solving Part 4: Aim at It "Connie," says Peter, "do you know which radio ads did bring customers? If we knew that, we might be able to buy just a few radio ads. We might get the same income and have some money left to put into other ads."

"I asked my staff this morning to find that out," says Connie. "I have the papers here. Let's see what they say. It looks as though the only ads that worked were around the financial news at 6 A.M. They cost $8,580 to run one ad every week. Those were the only ones customers mentioned hearing."

"Well," says Maria, "if you cut the rest of the radio ads, that's $17,160 − $8,580 = $8,580. Exactly half this year's radio budget! That's $8,580 you can take out of next year's budget and still keep 52 radio ads. We know you have to cut this year's budget by $10,332. That's $10,332 − $8,580 = $1,752. You still have to find some other ways to save $1,752."

USING CIRCLE OR PIE GRAPHS

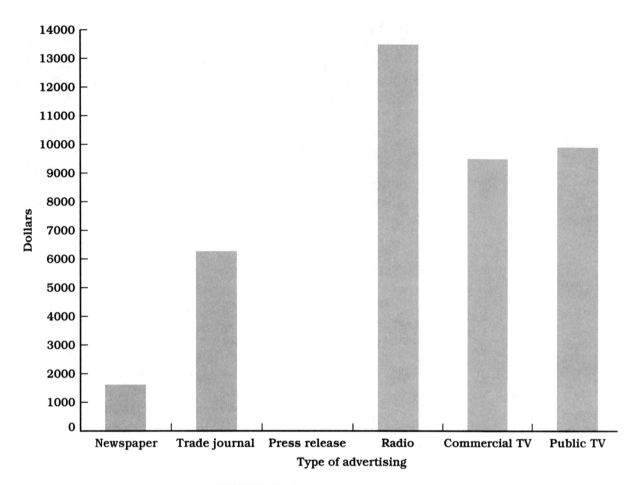

FIGURE 8.10
Total costs for next year at this year's costs less 20 percent

	This Year	Next Year
Newspaper	10	8.0
Trade journal	20	16.0
Press release	6	4.8
Radio	104	83.2
Commercial TV	4	3.2
Public TV	50	40.0
Total	194	155.2

FIGURE 8.11
Number of ads Connie can buy

"Yes, but where do I take the other money from?" asks Connie.

"If we did one less commercial TV ad, that would save $3,000 right there," Roy offers. "I know that could cost a lot in lost income, too."

"These decisions are never easy," adds Terry. "I would say cut more from radio. If you look at the total income figures, radio still brings in the least."

"I think you're on the right track," says Connie. "Look at this: If we only sponsored the financial news partly, say once a month, that would cost $1,980. It would still keep our name in front of regular listeners."

"Including our president, who never misses a broadcast," says Roy with a smile.

"OK, here's what it looks like now," says Connie. She shows them figures she's jotted down in Figure 8.12.

"Well, I can certainly go to George with these figures! Even if I buy another commercial TV ad, we'll still be within next year's budget of $41,328! We've met his request without losing a significant amount of income. If all goes well, I can make up any loss from radio with another TV ad and more press releases. They did bring in $30,000 last year from only six articles. Thanks, team! You really helped me set my priorities. I just hated to risk that $20,000 from radio. Now I really don't have to."

Problem Solving Part 5: Score It "We will all be interested to see how our predictions turn out. It would be great if income actually went up," says Roy. "Let's check the income we can trace to ads each month. That way we'll know how our plan is working."

	This Year's Cost	Next Year's Cost
Newspaper	$ 2,000	$ 2,000
Trade journal	8,000	8,000
Press release	0	0
Radio	17,160	1,980
Commercial TV	12,000	12,000
Public TV	12,500	12,500
Total	$51,660	$36,480

FIGURE 8.12
Proposed cuts to Connie's ad budget

USING CIRCLE OR PIE GRAPHS

SUMMARY You have watched Connie prepare data and begin the problem-solving process. She took her information to the team. The team members helped her clear up her thinking. Often others can help us problem-solve even when they're not experts in our area. Connie used both bar graphs and circle graphs to make it easy for the team to understand her data.

ACTIVITIES

1. Create your own circle graphs using a computer or by hand. Some topics that are interesting are:
    a. What percent of available sick days does your department use for each of the weeks in a month?
    b. What percent of the work day is spent making repairs to the line for the five days in a week?
    c. At home, you can figure and graph the percent of your month's budget that goes for food, rent, clothes, entertainment, etc. You might be surprised at what you find out.
    d. Pick a problem from your workplace. Make some bar or circle graphs to help you analyze your data. Use the problem-solving process to pick a possible solution. Present your information to the class.

Appendix A

Answers and Assistance

The correct or suggested answers for the Practices and Activities sections of each chapter are given below.

CHAPTER 1

Practices

Page 7
1. Belt thickness, strength — Roy
   Proper use of belts — Connie
   Cutter machines — Terry
2. 1. List all absences by day of the week
   2. Survey your workers
   3. Check cash register slips

Activities

1. Match the terms with the best meaning.

   a. team — a group of people working together with a common goal
   b. data — information that can be found and measured or counted
   c. population — the total number of people or things
   d. process — the way things are made, moved, or used, or the way people are served
   e. quality — the best possible condition
   f. brainstorm — a way to get many ideas quickly
   g. individual — one of a population of people or things
   h. cross-functional team — different teams, departments, or jobs working together to solve problems
   i. specification — a piece of information describing something

ANSWERS AND ASSISTANCE

2. a. (w) b. (w) c. (c) d. (w) e. (c)

3. a. machine operators—cutters—belts cut by each worker
   b. skilled trades—electricians—repair jobs completed
   c. plant fire extinguishers—hand-held tanks—gallons sprayed per minute
   d. sewing machines—Dressmaker Deluxe—size of belt
   e. managers—Roy Urumatsu—how many people supervised

4. Answers to this question will vary. You might have included people from the departments in the organization where you work. People on cross-functional teams do different jobs. They may or may not work in one department. Examples: Workers from shipping might include packers, drivers, and order input clerks. Another team might be machine operators, repair persons, receiving clerks, and inspectors.

5. Answers will vary. Cross-functional teams can work on any problem that makes a difference to the work of any team member. All the team members do not have to be responsible for actually doing the work or solving the problem.

CHAPTER 2 Practices

1. Page 16

	Monday	Tuesday	Wednesday	Thursday	Friday	Total
Week 1	10	13	15	9	11	58
Week 2	17	9	15	13	16	64
Week 3	25	10	3	6	19	63
Total	46	32	33	28	46	185

a. Monday and Friday
b. 46, 32, 33, 28, 46, 185

2. Page 17

Shift	Veteran's Day	Columbus Day	Total
A	40	30	70
B	20	50	70
C	15	55	70
Total	75	135	210

a. 70
b. Columbus Day — many more people wanted it.

3. Page 18

1. The range is Monday/Friday 46 – 28 = 18
   The sample period is the week.
2. d. 46, 32, 33, 28, 46 — more at the beginning and end

■ 104 ■

ANSWERS AND ASSISTANCE

Activities

1. **a.** highest: 41; lowest: 12; range: (41 − 12) = 29
   b. highest: 49,750; lowest: 31,200; range: (49,750 − 31,200) = 18,550
   c. highest: 9; lowest: 2; range: (9 − 2) = 7
   d. highest: 40,475; lowest: 38,900; range: (40,475 − 38,900) = 1,575

2. Answers could include anything safety related: clear doorways, fire extinguishers, hazardous materials storage, electrical close-outs, etc.

3. **a.** Groups such as customers, kinds of calls, equipment, and staff problems.
   b. Answers could include: staff, customers, kinds of calls, and equipment
   c. Your charts should look as follows: (Totals are added to facilitate the analysis in question d.)

1) New-Product-Related Calls

Shift	Monday	Tuesday	Wednesday	Thursday	Friday	Total
A	3	5	7	6	2	23
B	3	4	5	6	1	19
C	0	1	2	3	1	7
Total	6	10	14	15	4	49

2) Calls Longer Than 3 Minutes

Shift	Monday	Tuesday	Wednesday	Thursday	Friday	Total
A	6	8	10	6	2	32
B	5	6	8	10	1	30
C	3	4	6	8	2	23
Total	14	18	24	24	5	85

3) Calls Sent to Customer Service By Mistake

Shift	Monday	Tuesday	Wednesday	Thursday	Friday	Total
A	1	3	3	4	3	14
B	6	10	12	12	10	50
C	0	0	1	1	0	2
Total	7	13	16	17	13	66

   d. 1) Wednesday and Thursday
   2) 200
   3) B; C
   e. Answers will vary. Some possible responses may include: operator error, untrained operator, etc.
   f. Answers will vary. Some possible responses may include: train the operator, get a new operator, etc.

4. Answers will vary.

ANSWERS AND ASSISTANCE

CHAPTER 3 Page 33

Practices Compare your results with the tally chart on page 39 (Figure 4.2).

Activities
1. a. highest: 15.0; lowest: 7.5; range: 7.5
   b. 12.25

2. population; individual; tally; statistics; coincidence; cells; boundaries; interval; midpoint; average.

Connie wants to find out if her advertising is bringing new customers to Best Belt. She decides to analyze the phone calls coming in to the sales office. The total number of phone calls is the <u>population</u>. Each telephone call is an <u>individual</u> in the sample. The new ad appears in the trade magazine on Monday. Connie asks her staff to make a record of all the calls they receive. She wants them to divide the calls they get into three categories:

1. regular customers who saw the ad
2. new customers who saw the ad
3. new customers who did not see the ad

She asks the staff to keep a <u>tally</u> for four days: Tuesday, Wednesday, Thursday, and Friday. Connie believes that these <u>statistics</u> about the sample will give her valuable information about the whole population. She wants to be sure that the relationship between new customers and the ad is not merely a <u>coincidence</u>.

The staff makes a tally sheet with six <u>cells</u>. The cell <u>boundaries</u> are 0, 5, 10, 15, 20, and 25. That means the cell <u>interval</u> is 5. To make the data clear, the numbers are tallied at the cell <u>midpoint</u> halfway between each boundary. Connie's staff finds the midpoints by getting the <u>average</u> of the cell boundaries.

3–5. Answers will vary.

CHAPTER 4
Activities
1. a. 6
   b. 13
   c. 13
   d. 13

2. yes. The mean, median, and mode are the same.

3. a. mean = 13; median = 13; mode = 13.
   b. Yes. The three measures of central tendency are the same.

4. a. mean = 5; median = 5; no clear mode.
   b. There is no central tendency.

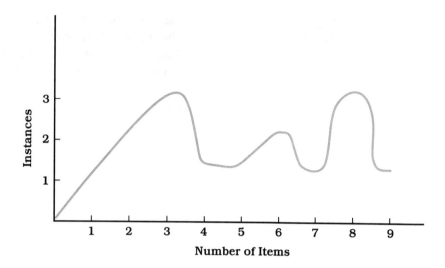

**CHAPTER 5
Activities**

1. Place the letter of the word next to its meaning:
   a. axis — one leg of a line graph
   b. standard deviation — the measure of dispersion
   c. stable process — always doing something in the same way
   d. dispersion — the spread of data in a distribution
   e. sigma — a Greek letter that stands for a standard deviation
   f. probability — the odds that something will happen
   g. vertical line — runs from top to bottom of a page
   h. horizontal line — runs across the page

2. a. Maria can't get copies of her report. (1)
   b. The copier needs toner. (2)
   c. This is the third time in six months the toner delivery has been late (1)
   d. Maria's secretary says the copier is not working. (1)
   e. Maria's secretary says, "Maybe one supplier is always late." (4)
   f. Maria asks why the copier is broken. (2)
   g. The secretary checks the storeroom. (2)
   h. The records are checked again, this time by supplier. (2)
   i. The monthly delivery records are checked. (2)
   j. Office Chief Company has had three late deliveries. (2, 3)
   k. Maria buys toner from three suppliers. (2, 3)
   l. All the deliveries are on time for the next six months. (5)

ANSWERS AND ASSISTANCE

   m. Maria calls Office Chief to ask that they be sure toner is delivered on time. (4)
   n. Records say the toner was ordered two weeks ago. (2)
   o. There is no toner in the storeroom. (3)

3. **a and b.**

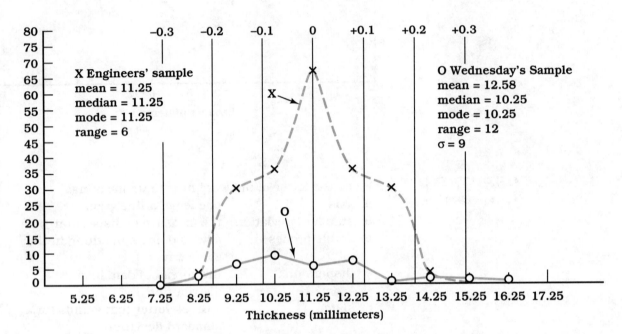

Wednesday's sample of 40 belts compared to the engineers' sample

   c. Different shaped curve, not bell shaped, wider dispersion, wider range, smaller sample size, etc.
   d. 8

4. Answers will vary.

5. 4.108974359

CHAPTER 6 Practices — Pages 66–67

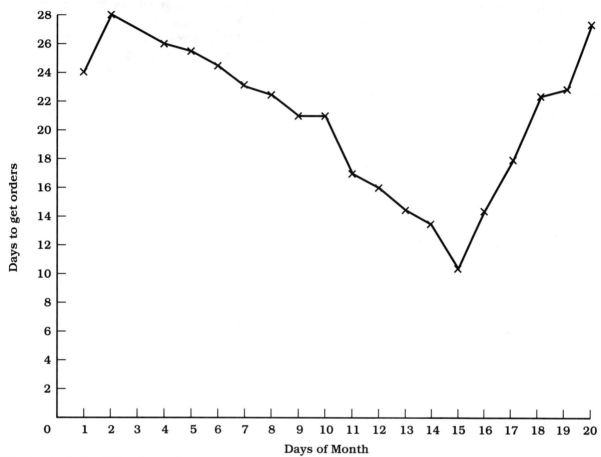

Means graph for 20 business days

MARIE'S DATA FOR 20 BUSINESS DAYS, WITH TOTALS AND AVERAGES

	Day 1	Day 2	Day 3	Day 4	Day 5	Day 6	Day 7	Day 8	Day 9	Day 10
Oil	26	30	26	27	26	25	23	24	24	18
Blades	22	26	28	25	25	24	23	21	18	24
Total	48	56	54	52	51	49	46	45	42	42
Average	24	28	27	26	25.5	24.5	23	22.5	21	21

	Day 11	Day 12	Day 13	Day 14	Day 15	Day 16	Day 17	Day 18	Day 19	Day 20
Oil	16	18	14	12	20	14	20	25	23	30
Blades	18	14	15	15	1	15	16	20	23	25
Total	34	32	29	27	21	29	36	45	46	55
Average	17	26	14.5	13.5	10.5	14.5	18	22.5	23	27.5

ANSWERS AND ASSISTANCE

Activities

1. Figure 6.8 on page 72 of the text shows the correct central limits.
    a. 14.5, 13.5, 10.5, 14.5. The mean is 13.25.
    b. 28, 27, 26, 25.5, 27.5. The mean is 26.8.
    c. Answers will vary.

2. a–d.

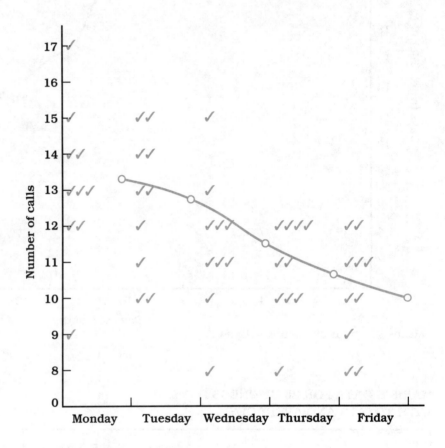

   e. The trend is downward.
   f. No, not long enough time.

3. Answers will vary.

**CHAPTER 7
Activities**

1. The three charts follow.

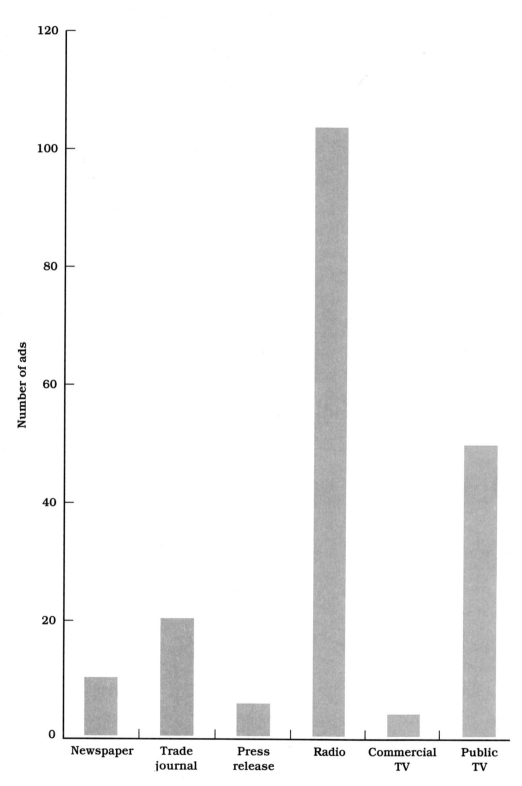

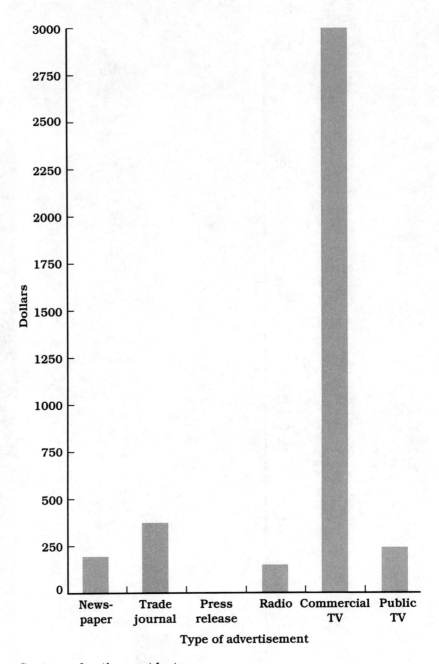

Cost per advertisement by type

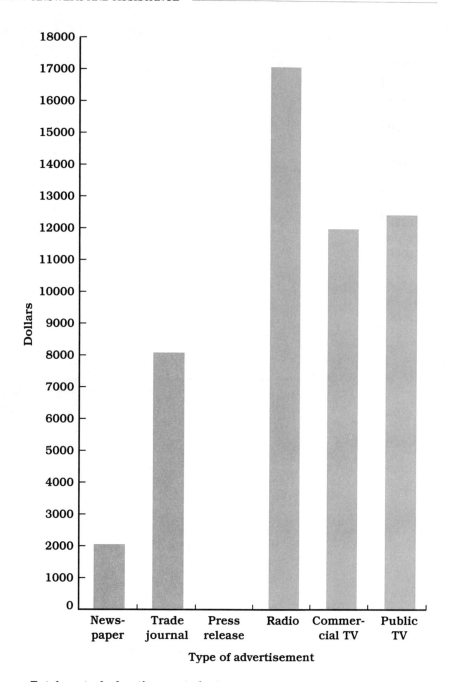

Total cost of advertisements by type

2.

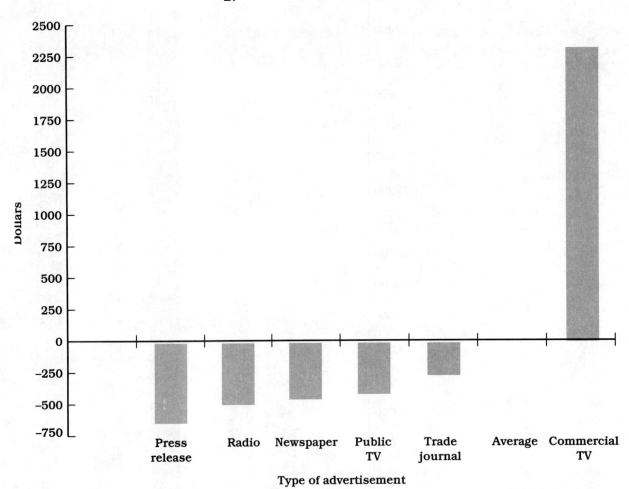

Cost per ad of each type in relation to the average

3. Answers will vary.

Chapter 8

1. There are no set answers for Chapter 8. It is recommended that you follow the procedure in activity 1 and share the results with someone to be sure you understand it.

Appendix B

Quick Math Review

WHOLE NUMBERS Whole number operations follow patterns. This section shows some ways to organize whole numbers.

Place Value Our whole number system works from right to left. The smaller (first) numbers are on the right. The largest value is on the left. So, 2,345,678 reads, in English: two million, three hundred forty-five thousand, six hundred seventy-eight. There are no "ands" in whole numbers. Notice that there are three digits between commas. Adding digits and commas to the left of a number increases its magnitude. For example, 4,982,345,678 reads: four billion, nine hundred eighty-two million, three hundred forty-five thousand, six hundred seventy-eight. Seventy-eight has seven tens and eight ones. Here's the pattern that occurs with groups of three digits separated by commas:

trillions			billions			millions			thousands			h	t	o
#	#	#,	#	#	#,	#	#	#,	#	#	#,	#	#	#
h	t	o	h	t	o	h	t	o	h	t	o	h	t	o
u	e	n	u	e	n	u	e	n	u	e	n	u	e	n
n	n	e	n	n	e	n	n	e	n	n	e	n	n	e
d	s	s	d	s	s	d	s	s	d	s	s	d	s	s
r			r			r			r			r		
e			e			e			e			e		
d			d			d			d			d		
s			s			s			s			s		

If you remember this pattern, you can read any number. Another useful pattern in carrying out arithmetical operations is called columns. Suppose you want to add three hundred forty-

■ 115 ■

five and seven hundred sixty-five. The pattern of columns looks like this:

thousands	hundreds	tens	ones
	1	1	
	3	4	5
+	7	6	5
1,	1	1	0

When you add, you may not put a tens place digit in the ones column, and so on. Now add the digits in each column, starting on the right. You know 5 + 5 = 10. What is 10? Right, it is zero ones and one ten. Put it on the chart. How? Put the 0 in the ones column under the 5s. Put the 1 (one ten) over the other tens in their column. Now add 1 + 4 + 6 = 11. (11 tens or one ten and one hundred. Repeat the pattern. Put the first 1 under the tens column, and "carry" the other 1 to the hundreds column. Add the hundreds: 1 + 3 + 7 = 11. Now what? Put the first 1 under the hundreds column and carry the other 1 to the next left column. This is a new column. Remember the reading numbers rule: three digits from the right, add a comma and name the group. What group will you be in when you put the last 1 in a new column? Right, the thousands group, the first (ones) column. Your answer is one thousand, one hundred ten. Sometimes, we call numbers like 1,110 eleven hundred ten. That's all right in speaking. However, you will be sure to write them correctly if you remember the pattern. To remember the comma rule—think: "Three strikes from the right—you're out!"

When you want to add, subtract, or multiply whole numbers, use the columns to keep things straight. Suppose you want to subtract 3,443 from 7,425. Subtract the same way you add:

thousands	hundreds	tens	ones
6	1 3	1	
7̶,	4̶	2	5
− 3,	4	4	3
3,	9	8	2

Again, work from the ones column: 5 − 3 = 2. Put the 2 in the ones column. Can you take 4 from 2? No. Now you have to "borrow" (the opposite of "carry"). Borrow one "hundred" from the hundreds column, leaving only 3 there. Now, you have 12 "tens." Take 4 away, and write 8 in the tens column. Can you take 4 from 3? No. Borrow again. Take one "thousand" from that column, leaving only 6, and make the 3 a 13.

Now subtract the hundreds column: 13 – 4 = 9. Can you take 3 from 6 in the thousands column? Yes. The answer is 3,982, or three thousand, nine hundred eighty-two. Notice that the comma appears in the written words as well as in the numbers.

Carry to the left; borrow from the left.

When you multiply, you carry in much the same way as in adding. Columns are very important when multiplying. Follow this pattern:

Add:	**Multiply:**
	1
345	345
+345	× 2
690	690

Multiplying is just a quick way to add. Why bother? Look at this problem:

Add:	**Multiply:**
	4 4
345	345
345	× 9
345	3,105
345	
345	
345	
345	
345	
+345	
3,105	

Which problem is easier to figure out? Once you have learned the multiplication facts, multiplying is much faster than adding. As you can see, multiplying 345 by 9 is the same as adding up 345 nine times. Notice the difference between adding and multiplying. If you add 345 and 15, the answer is 360. If you multiply 345 by 15, the answer is 5,175. (345 × 15 means adding up 345 15 times. 345 + 15 means adding 1 ten and 5 ones to 345.)

Here's a column pattern for multiplying large numbers:

				7,	4	2	5
			×	3,	4	4	3
			2	2	2	7	5
		2	9	7	0	0	*
	2	9	7	0	0	*	*
2	2	2	7	5	*	*	*
2	5,	5	6	4,	2	7	5

First, multiply all the top numbers by each digit in the bottom number, starting with the ones column. Carry as you would when adding, adding the "carry" digit to the result of that column's multiplication. Begin a new row each time you multiply the top number by a different digit of the bottom number. Each time you begin a new row, put the first (rightmost) digit of the answer in the column right below the digit you are multiplying by. You can use a star or a zero to mark the empty place or places to the right. Second, when you have multiplied all the digits, add each column as you would add any numbers. Place the commas by the rule: Three strikes from the right—you're out! The answer is 25,564,275. Read the number.

Dividing uses columns, too. When you divide, you use all your whole number skills: adding, subtracting, multiplying, and dividing. Just as we likened multiplication to adding the same thing over and over, we can liken division to subtracting the same thing over and over. For instance, 2,505 ÷ 50:

$$
\begin{array}{r}
2{,}505 \\
-50 \\
\hline
2{,}455 \\
-50 \\
\hline
2{,}405 \\
-50 \\
\hline
2{,}355
\end{array}
\quad \text{and so on.}
$$

Do you see a pattern? Multiplying is a fast way to add. Dividing is a fast way to subtract. It looks at how many sets of one number will "fit" into another number. How many sets of 5 will fit into 15? ***** ***** ***** = 15. Count the stars. Three sets of 5 will fit into 15; 15 divided by 5 is 3.

When you need to divide large numbers, making sets takes too long. Here's a pattern you can use:

$$
\begin{array}{r}
75{,}205 \\
375{\overline{\smash{)}28{,}201{,}875}} \\
-2625 \\
\hline
1951 \\
-1875 \\
\hline
768 \\
-750 \\
\hline
1875 \\
-1875 \\
\hline
\end{array}
$$

When you divide, you do the left numbers first. Look at the divisor (the outside number), 375. It has three digits. Look at the first three digits in the dividend (the inside number). Will 375 fit into 282? No. That means you must go over one more place and work with 2820. How many 375s can fit into 2820? To get a good estimate to try, round the numbers to 400 and 2800. Right, 400 will go into 2800 7 times. Write the 7 above the last digit of the set we used, 2820. Now multiply that 7 times 375 and put the answer, 2625, under the 2820. Subtract, and you get 195. Next, "bring down" the next digit of the dividend, 1, and you have 1951 to work with. How many times will 375 go into 1951? Rounding to 400 into 2000, you will try 5. Write 5 above the brought down digit and multiply. Be sure to check the results of the multiplication to make sure it's not too high or too low (meaning one more division could "fit" in that dividend). Subtract and continue the pattern. What if the divisor won't fit when you bring down the next digit? Write a 0 above the brought down digit. Bring down another digit and try again. Continue until all digits are brought down and you have done the last "fitting," multiplying, and subtracting. If some is left over at the end it is called a "remainder." In this case we came out even.

Here are the steps for dividing large numbers:

Count "out" (count the digits in divisor).

Count "in" (count the same number of digits in the dividend).

Fit it in (guess how many times it will fit).

Times it out (multiply your guess and put the answer under the dividend).

Subtract it out (subtract your "times" from the dividend).

Bring it down (bring down the next digit).

Begin again and times it out. Remember: Count out, count in, fit in, times out, subtract out, bring down.

Signed Numbers

Some whole numbers are positive and some may be negative. Any number without a sign in front of it is positive. Most numbers we deal with are positive. The most common negative number you see is probably on a thermometer. Do you live where the temperature gets below zero? That's a negative number. The weather forecaster says, "It's minus ten today." That means it's ten degrees below zero. Another use of positive and negative numbers is in business. A business that is losing money has greater expenses than income. Its balance is below zero, or negative.

Negative numbers are a way of talking about quantities that are less than zero. If my income is $7,000 and expenses are $8,000, my balance is negative. Not only am I "broke," I'm in debt. I still owe $1,000 and my paycheck is gone. When you work with signed numbers, think about the thermometer and the balance sheet.

One way to "see" positive and negative numbers is to use a number line. Read the line below:

$-10\ -9\ -8\ -7\ -6\ -5\ -4\ -3\ -2\ -1,\ 0\ +1\ +2\ +3\ +4\ +5\ +6\ +7\ +8\ +9\ +10$

When you add signed numbers, it's called "combining." When you combine signed numbers, you move either left or right on the number line. If –3 is added to (combined with) +5, what will you get? Find –3 on the number line. When you add a positive, move to the right (–2, –1, 0, +1, +2) Five places to the right puts you at +2.

Combine +3 and –5. This time you start at +3 and move left five places to end up at –2. Try combining –3 and –5. Did you get –8? You will notice that the sign of the answer in combining is always the same as the sign of the larger of the two numbers combined. Another way to work with signed numbers is to remove one from another. It's like subtracting. When you remove a signed number, you in effect change its sign and combine it. Let's try removing +3 from –6: (–6) – (+3) = what? "Walk" the number line. Start at –6, take away 3 positives. When you take away positives, you add negatives (move away from the positive side). Where did you end up? At –9. Try removing –3 from –6: (–6) – (–3) = what? Start at –6 and remove 3 negatives (or add 3 positives). You should come out at –3. You have moved toward the positive side. Subtracting a negative is like adding a positive (two minuses equals a plus).

The rules are:
For combining (adding)

$$(+) + (+) = +$$
$$(-) + (-) = -$$
$$(\text{Big } +) + (\text{small } -) = +$$
$$(\text{Big } -) + (\text{small } +) = -$$

For removing (subtracting):

Remove a + = add a –: (+3) – (+2) = +1 or (+3) + (–2) = +1 or 3 – 2 = 1
Remove a – = add a +: (+3) – (–2) = +5 or (+3) + (+2) = +5 or 3 + 2 = 5

Two minuses equals a plus: 6 − (−4) = 6 + 4

For multiplying and dividing:

+ with + = +
− with − = +
− with + = −
+ with − = −

On the number line:

Move right to add a positive or remove a negative.
Move left to add a negative or remove a positive.

MEASURING IN METRICS

The metric system is a way of measuring based on units of ten. Distance is measured in meters. Liters measure volume (liquid), and grams measure weight. Decimal points are often used in metrics because decimals are also based on ten.

The belts for the floor polisher were measured in millimeters. "Milli" means *thousand*. It takes 1,000 millimeters to make one meter. Think of a meter as about as long as a yardstick plus the width of your hand (about 40 inches). In between meters and millimeters are decimeters and centimeters. Decimeters ("deci" means ten) are tenths of a meter. Think about decimeters as dimes: You need ten to make a whole (dollar or meter). Centimeters are hundredths of a meter, about 4/10 of an inch (a little less than half an inch). Think about centimeters as cents (pennies): You need one hundred to make a whole (dollar or meter). Figure B.1 is a chart of metric measures for distance (meters), volume

Prefix	kilo	hecto	deka		deci	centi	milli
Value	Basic Unit × 1000	Basic Unit × 100	Basic Unit × 10	Basic Unit	Basic Unit × 0.1	Basic Unit × 0.01	Basic Unit × 0.001
Length	kilometer (km)	hectometer (hm)	dekameter (dam)	meter (m)	decimeter (dm)	centimeter (cm)	millimeter (mm)
Liquid	kiloliter (kL)	hectoliter (hL)	dekaliter (daL)	liter (L)	deciliter (dL)	centiliter (cL)	milliliter (mL)
Weight	kilogram (kg)	hectogram (hg)	dekagram (dag)	gram (g)	decigram (dg)	centigram (cg)	milligram (mg)

FIGURE B.1
Metric measures

(liters), and weight (grams). Have you bought food or other products measured in metrics?

Most countries use metrics to measure everything. The United States is using metrics now for many purposes. You will soon see them everywhere. Measuring in metrics is easy. There are no difficult fractions to work with; decimals are much easier to perform mathematical operations with. To go from one unit of measure to another, just multiply or divide by ten, one hundred, or one thousand. All you really need to do is move the decimal point—to the right for multiplying, to the left for dividing. In a way, a decimal point is a multiplier or divider by tens, hundreds, thousands, and so on. If you have fabric that is one meter wide, you can cut 10 belts that are one decimeter wide (1 divided by .1 = 10). You could cut 100 belts that are 1 centimeter wide (1 divided by .01 = 100). To find how many millimeters there are in 2 meters, use your calculator to multiply 2 by 1,000. That's right, 2,000. The decimal point just moves to the right three places, filling in with zeros: 2 becomes 2,000. Look at the chart. When you change from a larger unit to a smaller one, the decimal point moves right. When you change from a smaller unit to a larger one, the decimal point moves left.

See the conversions in Figure B.2. For each unit you "go by," move one place. To go directly from kilometers to meters, move the decimal two places. You "go by" hectometers. You see, as you go from larger to smaller units, the numbers get larger—a zero is added to the right side of the number. That is the same as moving the decimal point one place to the right. If you have 3.4 meters and convert to decimeters, move the decimal one place to the right (3.4 meters = 34 decimeters = 340 centimeters = 3,400 millimeters). Try it yourself. Write any number of meters on a sheet of paper. By moving the decimal point and adding or taking away zeros, convert up and down from unit to unit. Skip some in between, but remember you have to "go by" them. Use the chart to help you decide what to do.

Kilometers		Hectometers		Meters
7543.	=	75430.	=	754300.
7543.			=	754300.
7.5	=	75.	=	750.

FIGURE B.2
Converting between units of length

QUICK MATH REVIEW

FRACTIONS If you have trouble understanding fractions, try this: In $\frac{7}{8}$, there is one pie cut into 8 pieces. You have 7 of them, and the dog ate the other one. The bottom number (denominator) is always how many pieces the whole thing is cut into. The top number (numerator) is how many of them you have. That's why you can say $\frac{14}{16}$ is the same as $\frac{7}{8}$. When you have 7 out of 8 possible pieces, you have the same amount of pie as if the 8 pieces were originally cut in half and you got 14 and the dog got 2.

When you are adding and subtracting fractions, the denominators have to be the same. They can't perform together until they can speak the same language. For example, you can not add $\frac{7}{8}$ and $\frac{3}{16}$—they are not on speaking terms. You must change your eighths to sixteenths, which means you'll have fourteen of them—$\frac{14}{16}$. Now the two fractions can be added. See the operations below:

$$\frac{7}{8} \times \frac{2}{2} = \frac{14}{16} + \frac{3}{16} = \frac{17}{16}$$

If we had some not-so-nice fractions to add like $\frac{3}{4}$ and $\frac{1}{3}$, we can't make both denominators either 4 or 3. We need to look for a common denominator—a number they're each part of. What is the smallest number you can think of into which both 3 and 4 will go? Will 6 work? No. 8? No. How about 12—yes! To make $\frac{3}{4}$ into a fraction with a denominator of 12, the bottom number (4) was multiplied by 3, so you must also multiply the top by 3 ($3 \times 3 = 9$). What would you do to the $\frac{1}{3}$? Right, multiply top and bottom by 4. The two resulting fractions can now be added. The operations are as follows:

$$\frac{3}{4} \times \frac{3}{3} = \frac{9}{12}$$

$$\frac{1}{3} \times \frac{4}{4} = \frac{4}{12}$$

$$+\underline{} \text{ then add}$$

$$= \frac{13}{12}$$

Now, look what happened! We have 13 pieces of a 12-piece pie. That's what is called an *improper fraction*. It's top-heavy, and we can't allow that to stay—it's just not proper!

Actually, if our pie is cut into 12 pieces, and we've got all 12, we've got a whole pie, haven't we? And we've got a piece ($\frac{1}{12}$) left over from another pie. Now we write a *mixed number*, which is writing how many whole things you have and how many parts of another are left over. Our mixed number will be $1\frac{1}{12}$.

Subtract fractions the same way: Find a common denominator for the fractions, then subtract the numerators. If you are working with mixed numbers, you may run into a new problems. Try this: $3\frac{1}{3} - \frac{2}{3}$. The $\frac{1}{3}$ is not large enough to subtract the $\frac{2}{3}$ from it. We're in luck, however, because we have a whole number to borrow from. Since we know that 3 pieces of a 3-piece pie is a whole pie, we can call one whole pie by its fraction name: $\frac{3}{3}$. We can then add those $\frac{3}{3}$ to the $\frac{1}{3}$ we already have (don't forget to deduct that whole from the 3 whole pies). Now, we have 2 and $\frac{4}{3}$ pies, and we can easily take $\frac{2}{3}$ of one pie from that. (I hope that's not all for the dog!)

$$3\frac{1}{3} = 2\frac{3}{3} + \frac{1}{3} = 2\frac{4}{3}, \text{ then subtract}$$
$$2\frac{4}{3} - \frac{2}{3} = 2\frac{2}{3}$$

When you multiply fractions, you multiply the numerators together and the denominators together. It is not necessary to have the same denominators.

$$\frac{3}{4} \times \frac{1}{3} = \frac{3}{12}$$

Now, when we get $\frac{3}{12}$, we're not quite finished. We need to simplify the fraction.

Mathematics is always trying to simplify things. (Do you really believe that?) So, we need to reduce our fraction to *lowest terms*—that is, the lowest form of that fraction. Remember that we found that $\frac{7}{8}$ and $\frac{14}{16}$ are the same portions of the pie, but $\frac{7}{8}$ is in lowest terms and $\frac{14}{16}$ is not. A fraction is in lowest terms if there is no number that both the numerator and the denominator can be divided by. Try dividing them both by 2, 3, 5, or 7. The chances are, if those numbers don't work, it can't be done, and the fraction is already in lowest terms. In our fraction $\frac{3}{12}$, you can see that both 3 and 12 can be divided by 3.

$$\frac{3}{12} \div \frac{3}{3} = \frac{1}{4}$$

In the problem $1\frac{3}{4} \times \frac{1}{3}$, we have a mixed number to multiply by a fraction. In order to do that, we first have to return the mixed number to its improper fraction stage. Turn the whole number into a fraction with the same denominator as the fraction, and add them together.

$$1\tfrac{3}{4} = \tfrac{4}{4} + \tfrac{3}{4} = \tfrac{7}{4}$$

A quick and easy way is to multiply the whole number by the denominator, add the numerator, and there you have the numerator of your improper fraction. Say 1 × 4 = 4. Then add the 3 to get 7. Place the 7 over the 4, and the improper fraction is $\tfrac{7}{4}$. What you're doing is counting fourths of pies: 4 fourths in one whole pie and 3 out of 4 in another to make 7 fourths. Now you're ready to multiply:

$$\tfrac{7}{4} \times \tfrac{1}{3} = \tfrac{7}{12}$$

How about dividing fractions? Let's try this one: $1\tfrac{3}{4} \div \tfrac{1}{3}$. First we must make an improper fraction out of the mixed number.

Then, we're ready to divide, right? Wrong! When you divide with fractions there's a wrinkle in the process. Rather than dividing by a fraction, what you do instead is multiply by the *reciprocal.* (That's good news for those of us who hate division!) The reciprocal is the fraction flipped over. Remember, it's always the second number that gets flipped, so be sure you keep them in order. The operations are as follows:

$$1\tfrac{3}{4} \div \tfrac{1}{3} = \tfrac{7}{4} \div \tfrac{1}{3} = \tfrac{7}{4} \times \tfrac{3}{1} = \tfrac{21}{4} = 5\tfrac{1}{4}$$

Notice that we got an improper fraction. Divide the 4 into the 21. It goes 5 times and 1 is left over. Put it on top of the 4, and you're home free!

Remember the rules for fractions:

1. Don't leave a fraction as an improper fraction.
2. Always reduce a fraction to its lowest terms when you're done.

DECIMALS AND PERCENTS

Many of the problems students have with decimals are caused by not understanding how they relate to fractions. Figure B.3 helps you look at decimals as fractions and understand "place value." Note that as you move farther to the right of the decimal point, the denominators *increase* and the value *decreases.* If you have $\tfrac{3}{10}$ (.3) of a pie, you have a lot more than you would if you have $\tfrac{3}{100}$ (.03), right?

An easy way to change any decimal to a fraction is to count the places to the right of the decimal point, place the same number of zeros below the line, and place a 1 in front of

Term	Fraction	Decimal
three-tenths	$\dfrac{3}{10}$	.3
three-hundredths	$\dfrac{3}{100}$	.03
three-thousandths	$\dfrac{3}{1,000}$	.003
three ten-thousandths	$\dfrac{3}{10,000}$	.0003
three hundred-thousandths	$\dfrac{3}{100,000}$	.00003
three-millionths	$\dfrac{3}{1,000,000}$	.000003

FIGURE B.3
Fraction and decimal equivalents

the zeros as the denominator. Then remove the decimal point and any beginning zeros from the original number for the numerator. For example, look at .003; there are three places to the right of the decimal. Place three zeros under the line and put a 1 in front. The denominator is 1,000.

You remove the decimal point and zeros to the left of the numerator only after you've formed your fraction. Why? The .00 before the 3 is what tells us we're looking at three one thousandths. Since we've now put the one thousand as the denominator, we don't need the decimal point and zeros any longer. Study Figure B.3.

The way to change a fraction into a decimal is to divide the numerator by the denominator (the top by the bottom). You will need to add a decimal point and some zeros to the numerator first, then carry out the long division. For example, $\frac{1}{4} = 1 \div 4 =$

$$
\begin{array}{r}
.25 \\
4\overline{)1.00} \\
\underline{8} \\
20 \\
\underline{20}
\end{array}
$$

A less neat example is $\frac{1}{3} = 1 \div 3 =$

$$3\overline{)1.000}.333\ldots$$

```
      .333 . . .
   _____
3 ) 1.000
    9
    ─
    10
     9
     ─
     10
      9
      ─
      1
```

If you get a remainder after you've added a third zero, you've probably got a nasty fraction. Another way to tell is if, when you divide, each time you subtract, you keep coming up with the same remainder (1 in this case). Then you know your decimal will go on into infinity.

Practice turning fractions into decimals and percents, using what you have learned above. Once you have done several, you will find them easy to do.

Another place where people have trouble is changing percents to fractions or decimals, or the other way around. Some are easy to work with. For example, 50% is the same as .50 and the same as $\frac{1}{2}$. Think of .50 as 50 cents, and $\frac{1}{2}$ as a half-dollar. Unfortunately, not all "fractions to decimals to percents" are so cooperative. Some are downright nasty. If you use a calculator, you've probably had a problem come out something like 45.333333333 or however many places your machine can go. This is the famous .3333 . . . decimal, which equals $\frac{1}{3}$, or $33\frac{1}{3}$%. You can wrap those threes clear around the world beginning at your home, and they'll still be going when you get back. That's called infinity. It's no use trying to work a problem using those values in any other way than fractions, or with a calculator.

Look at the first line of the chart in Figure B.4. Some awkward percents, like $12\frac{1}{2}$%, will behave if you carry them to three places as decimals. That's a good idea. An even better idea is to do the operation in fractions (using $\frac{1}{8}$). As an example, find $12\frac{1}{2}$% of 225 using decimals.

$$12\frac{1}{2}\% = 12.5\% = .125$$

Now you can do the long multiplication to find .125 × 225, and you'll finally come to the answer, 28.125. You can find it more easily, if you're a fraction person, like this:

$$\frac{1}{8} \times \frac{225}{1} = \frac{225}{8} = 28\frac{1}{8}$$

Fraction	Decimal	Percent
$\frac{1}{8}$	.125	$12\frac{1}{2}\%$
$\frac{3}{8}$	.375	$37\frac{1}{2}\%$
$\frac{7}{8}$	.875	$87\frac{1}{2}\%$
$\frac{1}{6}$	.1666...	$16\frac{2}{3}\%$
$\frac{1}{3}$	.3333...	$33\frac{1}{3}\%$
$\frac{2}{3}$	.6666...	$66\frac{2}{3}\%$

NOTE: In most cases, you may round the equivalent of $\frac{1}{6}$ as a decimal from .1666 to .167, and as a percent (in decimal form) from 16.667% to 16.7%. You may also round $\frac{2}{3}$ as a decimal from .6666 to .667, or as a percent from 66.667% to 66.7%. Rounding, or estimating as it is sometimes called, will not make an important difference in your answer.

FIGURE B.4
Equivalent chart for uneven decimals, fractions, and percents

If you can memorize math facts, it may be easiest to just learn these equivalents, so you can use them quickly. Some are very easy and familiar because they translate into money terms. For example, $.25 = 25\% = \frac{1}{4}$ (think of a 'quarter" or 25 cents). Those are not on the chart. The more difficult ones are.

If you cannot memorize math facts easily, remember the following:

1. Percents can be made decimals by removing the percent sign and moving the decimal point two spaces to the left.
2. Decimals can be made percents by moving the decimal point two spaces where?—you guessed it—to the right and adding a percent sign. If you only had two decimal places to begin with, you'd not see the decimal point once you moved it. Examples: .25 = 25%, .62 = 62%, and 2.54 = 254%, but .254 = 25.4%.
3. To change a percent or decimal into a fraction, put the number (less the decimal point and beginning zeros) over the appropriate denominator as detailed earlier, then reduce to lowest terms:

$$25\% = .25 = \frac{25}{100} = \frac{1}{4}$$

4. To change a fraction into a decimal, divide the numerator by the denominator.

To solve problems using percents, follow the handy chart in Figure B.5.

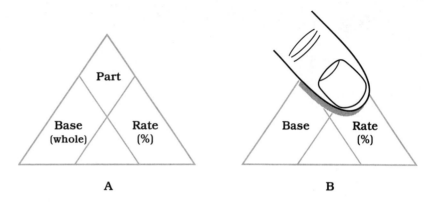

FIGURE B.5
Triangle A represents the Business Person's Rule (base, part, rate). In triangle B if you cover over the word "part", the formula reads "base × rate."

1. Notice that an X separates the % and the Base. To find the Part (amount of the base equal to that percent), multiply the base by the percent (sometimes called the rate). For example, 15% of 1200 equals what part?

    ```
    1200   Base
    × .15  % changed to a decimal
    6000
    1200*
    180.00  The part is 180.00 or 180.
    ```

2. To find the percent (rate), divide the part by the base. Notice that the chart shows Part/Base when you cover the % with your thumb. The / line means "divide."

    ```
                  .15 = 15%
    Total 1200 ) 180.00  Part
                 1200
                 6000
                 6000
    ```

3. To find the base, divide the part by the percent (rate): When you cover the Base, you see Part/Rate.

    ```
              1200.
    .15)180.00
         15
         30
         30
    ```

    Be sure to first move the decimal point in the divisor (the outside number) two places to the right to make a whole number and then move it an equal number of spaces in the dividend.

Glossary

average (**a** ver age) A middle point in a group of numbers. Averages are found by adding all of the sample numbers together and then dividing that total by the number of samples.

axis (**ax** iss) Each of the two lines in a line graph. (Note that the plural of axis is axes.) The lines are either vertical or horizontal.

bell curve (**bell kur** ve) A normal curve that is shaped like a bell; both sides are equal.

bimodal (**by** mow dal) Having two modes. On a graph, the distribution has two high points. There are fewer items at the mean (average) than at the high points.

boundary (**bown** dah ree) Limits to the size of a cell; its top and bottom.

brainstorm (**brain** storm) A way to get many ideas quickly. Team members offer ideas as fast as they can think of them. Every idea is okay and written down until there are no more. Ideas can be grouped and then discussed.

cell (sel) A group of data organized into a unit.

central tendency (**sen** trall **ten** den c) When a large number of measures is taken, most will be close to the middle of the range. A few will be at each end of the range.

circumference (sir **come** fur enss) The round edge of the circle. It can be measured in degrees and always has 360 degrees, no matter what size it is.

coincidence (co **in** si dense) Two things happening at the same time or in the same place by chance. For example, two people born in the same town have a place relationship. Two people with the same birthday have a time relationship. They did not do that on purpose, or cause the other person to live in that town or be born on that day. Their relationship is coincidental.

compass (**com** pass) A tool for drawing circles, with a sharp point on one arm and a pencil on the other. The two arms are joined at the top and crossed by a curved "ruler." The ruler determines the radius and thus the size of the circle.

cross-functional team (kross **funk** shun al) People who have jobs in different departments coming together to work on a plan or solve a problem. The problem might involve more than one department.

cycle (**sy** kl) A pattern that repeats itself over and over. For example, the seasons of the year and the temperatures that go with them.

data (**day** ta) Pieces of information about a product or process.

diameter (die **am** eter) The line that cuts a circle in two equal pieces. Circles are usually named by their diameter. A six-inch circle has a diameter of six inches.

dispersion (dis **per** zyun) How far the data spreads on the horizontal axis.

distribution (dis tri **bu** shun) How the things or events are arranged. For example, the manager might want to know how December sick days are distributed. He wants to know if more sick days are taken on Mondays and Fridays than on Wednesdays. A frequency distribution table will tell him two things: how many sick days were taken and where in the work week.

estimate (**s** ti mate) A number that is close enough to give good information but is not completely accurate.

frequency (**free** kwen see) How many things there are or how often something happens. For example, a manager might want to know the frequency of sick days taken in December.

horizontal (**hor** i zon tal) A line that runs from left to right, as does the "line" where the earth meets the sky—the horizon. So any line like that is called horizontal.

interval (**in** turr val) The amount of space, measured in numbers, between the sides (boundaries) of a cell. All the cells have the same size interval.

just-in-time (**or JIT**) When a company tries to have on site (in their warehouse) only as much as is needed to keep the process going, they have a just-in-time inventory.

GLOSSARY

mean (**mee** n) Average.

median (**mee** dee n) The middle value in a row of numbers. Half the measurements are less than the median and half are greater.

mode (**mo** de) The number that shows up most often in a sample. It does not have to be near the median, and it may move the average away from the median.

normal (**nor** mal) The way things happen most of the time. In a normal distribution, the number with the highest frequency is close to average.

outlier (**out** lie er) A single number that is much farther from the mean than any other number in the set. It may be at either end of the distribution. It stands alone far outside the normal range. It is often not counted in the average because it moves the average too far away from the median.

population (pop u **la** shun) A group of people or things that can be counted or measured. All the people at a ball game make up the population of the stadium.

probability (prah bab **il** i tee) The likelihood that something will happen. Another word commonly used is *odds*. In the United States, there is a greater probability of snow in January than in June. The odds are better that it will snow in January than in June.

process (**pro** ses) A way of doing something. Mixing ground meat with salt and pepper, cooking it, and putting it in a bun is a process for making hamburgers.

protractor (**pro** track tore) A tool which measures angles and circles in degrees. It looks like a half circle.

quality (kwa li tee) A state of excellence.

quality control (**kwa** li tee kon **trol**) To check that every product meets a certain degree of excellence.

radius (**ray** dee us) The distance or the line from the center of a circle to anywhere on the edge. The radius is one-half the diameter. In a circle graph, the space between the radius lines shows the shares.

random (**ran** dum) By chance; scattered all through the population with no pattern or system. There are mathematical rules for gathering random samples in statistics.

range (raynj) A number showing how much difference there is between the smallest and the largest sample measurement.

run chart A graph showing how a process works over time.

It may show upper and lower control limits. It will show when a process is out of control. It can help to find the cause.

run graphed data for a series of events that shows a pattern over time.

sample A group of people or things taken out of the whole population. The sample group must be like the population in important ways. The sample individuals might be the same size, age, or fabric as the population as a whole. Studying the sample can tell about the population without inspecting every individual.

scattergram (**skat** er gram) The picture shown by plotting raw data on a graph. It may show a left-right or up-down pattern, or no pattern.

significant (sig **nif** i cant) Important to what we are doing. A significant thing is a "sign" that we want to pay attention to.

skewed (**s** qued) Off center, out of balance, having more at one end than at the other.

specification (spes if i **ka** shun) Data that tell how an item should look or be. Size, strength, color, and material are just some pieces of information that might be called specifications or "specs."

standard deviation (**stand** r d dee vee **a** shun) A measure of distance from the mean. It is used to understand the dispersion of a sample.

statistics (sta **tis** tiks) Numbers found by looking at a small sample from a whole population. These numbers are then used to describe the larger group.

table A way of showing data. The numbers are put in columns and rows so they can be seen quickly and clearly.

tally (**tah** lee) A way of counting by making a mark for each person, thing, or event. Often a special tally sheet is used. It may have cells already on it.

tolerances (**tol** er n s) The acceptable limits; not exact but close enough. In every sample, some individuals are "okay." They are not too large or too small to work well. Sometimes tolerance and specification are used to mean the same thing. They both mean within usable limits.

trend (tr **end**) A pattern of events that has meaning. It probably has an assignable (explainable) cause.

variation A change in a process from what is expected. Every process has some things that change while it is going on.

vertical (**ver** ti cal) A line that runs up and down.

Index

Average, 23, 30–33, 131
Axes, 47, 131

Bar graphs:
  to show distributions, 81–87
  overview, 77–81
Bell curve, 37, 44, 131
Bimodal, 77, 85, 131
Boundary, 28, 131
Brainstorming, 1, 5, 131

Cause and effect, 24
Cell, 28, 30, 131
Central tendency, 37, 40–42, 131
Circle graphs, 89–130
Circumference, 89, 93, 131
Coincidence, 23, 24, 131–132
Compass, 89, 91–92, 132
Cross-functional team, 1, 3–4, 132
Cycles, 63, 70–71, 132

Data, 1, 6–7, 26, 55–57, 132
Decimals, 125–129
Diameter, 89, 91–92, 132
Dispersion, 47, 49, 132
Distributions, 13, 18–19, 53–54, 132
  bimodal, 84–87
  skewed, 82–84

Estimates, 23, 30–31, 54, 132

Fractions, 123–125
Frequency, 13, 18–19, 26, 132

Horizontal axis, 47–48, 132

Individual, 34
Information gathering, 14–17
Interval, 23, 29, 71–72, 132

Just-in-time (JIT), 23, 24–25, 132

Line graphs:
  reading, 47–49
  using, 63–87

Mean, 37, 38–39, 133
Median, 37, 40, 41–42, 133
Metrics, 23, 26, 121–122
Midpoint, 34
Mode, 37, 40, 41, 42, 133

Normal, 37, 40, 133

Odds. *See* Probability
Outlier, 37, 42, 63, 73, 133

Percentages, 90–91, 125–129
Population, 1, 6–7, 133
Probability, 47, 52–54, 133
Problem solving:
  analyzing data, 17–19

INDEX

Problem solving (*cont.*)
  evaluating the change, 23–33
  finding the solution, 47–61
  gathering data, 14–17
  predicting the cause, 19
  problem evaluation, 5–8
  process, 13–14, 47–59
Process, 1, 4, 133
Protractor, 89, 91–92, 133

Quality, 1, 4, 133
Quality control, 13, 14, 133

Radius, 89, 91–92, 133
Random, 23, 25–26, 133
Range, 13, 17, 133
Run, 63, 73, 134
Run charts, 63, 70, 71, 72–75, 134

Sampling, 13, 14–15, 25, 134
Scale, 68–69, 71–72
Scattergrams, 63, 65–67, 134
Signed numbers, 119–121
Significant, 13, 17, 134
Skewed, 77, 82, 134
Specification, 1, 8, 134
Spread. *See* Dispersion, 49
Standard deviation, 47, 52, 134
Statistics, 23, 24, 134

Table, 13, 15–17, 134
Tally, 23, 26–28, 134
Teams, 1, 3–4
Tolerances, 37–40, 134
Trend, 63, 66–69, 134

Variation, 63, 71, 134
Vertical axis, 47, 48, 134

Whole numbers, 115